Leading the Sustainable Organization

This book is the first to combine the much talked about topics of leadership and sustainability, and provides readers with a comprehensive overview and pragmatic approach to leading sustainable organizations.

Chapters include discussions, case examples, steps, and useful tools centered on the components of the Leading the Sustainable Organization model. This model provides managers with a pragmatic, end-to-end framework for creating (in the case of new entities) or shifting (in the case of existing firms) their organizations' workforces to a sustainability focus.

Leading the Sustainable Organization is the perfect tool for executives and managers in small, medium, and large companies, and in all industries, to assist with the difficult and confusing topic of leading sustainability efforts.

This book will be of great interest to students and academics who want to learn more about corporate sustainability.

Timothy J. Galpin is an Assistant Professor at Colorado Mesa University. He has over 20 years' experience as a management consultant and business manager, working with firms around the world on strategic planning, strategy execution, merger and acquisition integration, restructuring, and organizational change.

J. Lee Whittington is Professor of Management at the University of Dallas. He focuses his teaching, research, and consulting in the areas of leadership, organizational behavior, and spiritual leadership.

R. Greg Bell is an Assistant Professor at the University of Dallas where he teaches courses in Strategic Management and Business and Society. His research focus is in corporate governance and sustainability.

Leading the Sustainable Organization

Development, Implementation and Assessment

Timothy J. Galpin, J. Lee Whittington, and R. Greg Bell

Routledge
Taylor & Francis Group

LONDON AND NEW YORK

First published 2012
by Routledge
2 Park Square, Milton Park, Abingdon, Oxon, OX14 4RN

Simultaneously published in the USA and Canada
by Routledge
711 Third Avenue, New York, NY 10017

Routledge is an imprint of the Taylor & Francis Group, an informa business

British Library Cataloguing in Publication Data
A catalogue record for this book is available from the British Library

Library of Congress Cataloging in Publication Data
Galpin, Timothy J., 1961-
Leading the sustainable organization : development, implementation, and
assessment / Tim Galpin, J. Lee Whittington, and Greg Bell.
p. cm.
1. Leadership. 2. Strategic planning. 3. Organizational effectiveness.
I. Whittington, J. Lee. II. Bell, Greg. III. Title.
HD57.7.G346 2012
658.4'092--dc23

ISBN: 978-1-84971-466-2 (hbk)
ISBN: 978-0-415-69783-5 (pbk)
ISBN: 978-0-203-12844-1 (ebk)

Typeset in Times New Roman
by Saxon Graphics Ltd, Derby

Printed and bound in Great Britain by
TJ International Ltd, Padstow, Cornwall

Contents

List of Figures

Acknowledgments

We would like to acknowledge our clients and colleagues, who are too numerous to name here, but who have added much to our understanding of what leading successful sustainability efforts entails. We would also like to thank Lucy Gonzalez at Mission Foods, Tom Kemper at Dolphin Blue, Scott Morrison at City Garage, and Phillip Williams at Montgomery Farm for the generous time they took to provide us with their insights and experiences in leading their sustainable organizations.

Preface

As sustainability has grown in popularity, the availability of books and articles on the topic has also increased. Whether identified as sustainability, corporate social responsibility (CSR), corporate social performance (CSP), "going green," or the "triple bottom line" – including financial, environmental and social performance – the public's interest in the area is far-reaching.

Just as sustainability is still ill-defined, the writing thus far on the topic is also scattered and fragmented. What has been written on sustainability falls, for the most part, squarely into the "thou shalt" genre of management books. That is, the authors who address the topic of corporate sustainability (for the purposes of this book, we use the terms corporate sustainability, CSR, CSP, "going green," and the "triple bottom line" interchangeably) seem to all agree that executives and managers should be doing something about making their organizations more sustainable. However, there is limited, practical guidance regarding how to go about leading and engaging a sustainable organization. While there are numerous anecdotes and case examples of sustainability activities that companies have engaged in, there is no clear end-to-end model to aid executives and managers in leading sustainable organizations. What is needed is a comprehensive and pragmatic framework that addresses sustainability from strategy, through employee engagement, to measurable results.

Why this book?

This is the first book to combine leadership with sustainability. In view of the scattered nature of information, and more importantly, useful how-to information on the topics of leadership and sustainability, we provide readers with a comprehensive overview and pragmatic approach to leading sustainable organizations. The chapters of this book include discussions, case examples, steps, and useful tools centered on the components of our Leading the Sustainable Organization model. This model provides managers with an evidence-based, pragmatic, end-to-end framework for creating (in the case of new entities) or shifting (in the case of existing firms) their organizations' workforces to a sustainability focus.

This book is divided into three parts. Part I – Development – focuses on the macro-components of leading sustainable organizations taking place at the firm-wide and divisional levels. Part II – Implementation – concentrates on the micro-components of the model occurring at the manager/supervisor to employee level. And Part III – Assessment – addresses the two levels of sustainability results that can be achieved, at the micro-/employee level and at the macro-/firm-wide level. The chapters of *Leading the Sustainable Organization* include:

- an introduction to the Leading the Sustainable Organization model; an overview of the current state of corporate sustainability; examples of recent industry efforts; a summary of sustainability research findings to date; and various definitions of sustainability.
- Part I – Development: the macro-components of leading sustainable organizations – describes how sustainability is integrated into an organization's mission, values, goals, and strategy, along with how an organization's HR value chain reinforces the firm's sustainability efforts.
- Part II – Implementation: the micro-components of leading the sustainable organization – presents the "full-range leadership" approach to engaging employees in a firm's sustainability efforts.
- Part III – Assessment: the outcomes of leading the sustainable organization – explains both the micro-level/employee and the macro-level/firm-wide sustainability performance that can be achieved.

The book's summary includes a discussion of sustainability "killer phrases," the seven deadly sins of sustainability efforts, and the key success factors for leading a sustainable organization.

Each chapter also includes: a discussion of the concepts identified in the model; clear case examples illustrating the concepts discussed and how those concepts are being implemented by companies, including Ford Motor Company, Seventh Generation, Ecolab, Patagonia, Marks & Spencer, Whole Foods, Unilever, Henkel, and Siemens; tools and templates that can be used within organizations as practical aids to management; 'key take-aways' that readers should stay mindful of during the pursuit of their sustainability efforts; and discussion questions for use by instructors of sustainability courses in undergraduate and MBA programs.

Who should read this book?

Leading the Sustainable Organization was conceived as both a pragmatic, how-to handbook for executives, line managers, and leaders of support functions (including HR, finance, IT, and so forth) to assist them with the difficult and confusing topic of leading sustainability efforts, as well as a book that can be used by academics in their teaching and research on corporate sustainability. Because a firm's sustainability efforts can be organization-wide, divisional, or even departmental in scope, this book can be used by senior executives for organization-wide efforts, as well as by middle management and supervisors for sub-organizational sustainability implementation within their divisions or departments, in both line and support functions.

Moreover, because the Leading the Sustainable Organization model was developed from a multi-level, multi-disciplinary perspective, this book can be used in classrooms to provide students with an over-arching view of sustainability, from strategy, to workforce engagement, to results. Because the framework is based on sustainability research and evidence from various disciplines, researchers can use the model presented to frame focused research in the broader context of over-arching, multi-disciplinary sustainability efforts.

Throughout the text are numerous examples of companies and their sustainability strategies, implementation efforts, and achievements. In organizations, as in any other constantly evolving environment, information and understanding change. Consequently, the case examples provided in this book are time-based. That is to say, a company identified as being successful with its sustainability efforts today can become a laggard in corporate sustainability tomorrow. Conversely, the laggards of today can become the sustainability leaders of tomorrow. Therefore readers should view our examples as illustrative of the points being made, not as timeless gospel that will last forever. Such "business gospel" does not exist, especially in today's constantly and rapidly changing global business environment. Rather, our readers should take the specific company examples illustrated as good ideas that they should consider applying to their organization.

We have endeavored in *Leading the Sustainable Organization* to bridge the often-decried gap between academic theory and management practice. We have also attempted to break down a complex topic into its manageable components. This book is intended to be used as a how-to manual by organizational leaders who are either currently or potentially embarking on sustainability efforts, by instructors who are tasked with teaching sustainability, and by researchers attempting to gain a better understanding of what makes some corporate sustainability efforts more successful than others. We hope our readers agree that what they find within this book helps them to navigate through and effectively address the thorny issues that inevitably crop up during any sustainability effort.

Timothy J. Galpin,
J. Lee Whittington,
R. Greg Bell
July 2011
Dallas, Texas

1 Introduction

Sustainability has become quite fashionable over the past several years. Indeed, the "Sustainability Revolution" (coined by Andres Edwards, 2005) is creating a deep and enduring shift in people's consciousness and worldview. Sustainability now appears to be *the* strategic imperative of the new millennium. Today's authors use the phrases sustainability, corporate social responsibility (CSR), corporate social performance (CSP), going green, or the "triple bottom line" (Elkington, 1998; Savitz and Weber, 2006) to express society's desire to improve the long-term economic, social, and environmental performance of firms.

Yet, despite the ever expanding volume of literature underscoring the importance of sustainability to organizations, executives and managers do not have a clear roadmap identifying the full scope of factors affecting the success of sustainability initiatives. Certainly, much has been written regarding the strategic imperative of sustainability in helping firms achieve a range of organization-level outcomes. But, despite the increasing importance of sustainability, very little attention has been paid to the importance of individual managers and employees in helping identify, champion, coordinate, and implement sustainability efforts within organizations. There is little guidance on the role of leaders in helping foster and nurture an organizational environment that both understands and actively seeks sustainable solutions for the firm. Undoubtedly, it is important for leaders at the highest echelons of organizations to consider sustainability from a firm-level strategic perspective. However, the very best cost-saving ideas and innovative solutions that simultaneously promote social and environmental benefits will often come from front-line supervisors and employees. Unfortunately, there is little guidance available to company executives and managers to help them understand the critical interdependency of both management *and* employees in the success of sustainability strategies.

Leading the Sustainable Organization aims to fill this void. We see that the very success of sustainability strategies depends just as much upon the mobilization, eagerness, and enthusiasm of internal stakeholders of the firm as it does upon external stakeholders. In other words, management needs to recognize that the path to sustainability hinges upon a comprehensive understanding of both macro (organization-wide) and micro (manager-to-employee) factors associated with its success. The book is structured around the components of a model that

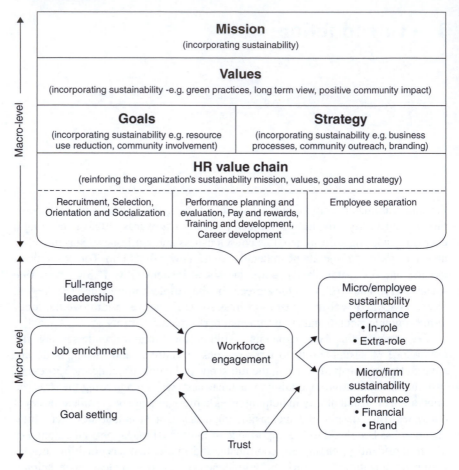

Figure 1.1 The 'Leading the Sustainable Organization' model

encompasses both of these dimensions, which we call the Leading the Sustainable Organization model (see Figure 1.1).

> **Key principle**
>
> The path to successful sustainability efforts hinges upon both macro- and micro-organizational factors.

Why the model was developed

In short, sustainability efforts have not been as productive as they could be. This is partly because management teams often fail to connect sustainability to business strategy (Porter and Kramer, 2006). Moreover, once a sustainability strategy has been set, firm leadership plays a crucial role in creating and maintaining effective strategic change. However, no clear leadership model exists with the express purpose of creating the most effective employee engagement during corporate sustainability efforts (Rok, 2009). In order to address these significant shortcomings, we believe there is a critical need for a comprehensive model that addresses both the macro- and micro-practices of successful sustainability efforts. Our multi-level model provides a comprehensive foundation for the formulation, implementation, and successful execution of sustainability efforts.

Key principle

There is a clear need for a leadership model that facilitates employee engagement in a firm's sustainability efforts.

How the model was developed

The Leading the Sustainable Organization model presented in Figure 1.1 was developed using a combination of the available practitioner and academic literature from multiple fields, including: sustainability, strategy, leadership, organizational culture, human resources, organizational behavior, and workforce engagement. This material included both qualitative and quantitative information. The formation of the model was also augmented by the authors' combined experience as seasoned practitioners and consultants in the areas of strategy development and execution, organizational culture change, and leadership development.

Synthesis rather than analysis

Because the model is meant to be used by managers and academicians alike, a systematic, statistical meta-analysis, as traditionally used by researchers, was determined not to be appropriate for its development. Moreover, meta-analysis has been contested as a means of understanding multi-faceted management endeavors such as sustainability efforts. First, meta-analysis rejects qualitative and anecdotal information that is often important in the development of complex knowledge (Denyer and Tranfield, 2006). Second, meta-analysis lacks an ability to cope with the variations in study designs, populations, contexts, and types of analyses (Cook *et al.*, 1997) that are found in a fragmented field such as management. Third, combining studies to derive an average removes critical contextual information (Hammersley, 2001).

In view of these limitations, the contribution of qualitative information has recently been recognized and has become progressively more valued as an evidence-based approach to building actionable knowledge in the field of management (Denyer and Tranfield, 2006). Therefore, due to the cross-functional and complex nature of the undertaking from which our model is derived (corporate sustainability efforts), the method used to develop it was a *narrative synthesis* of both practitioner and academic literature from various disciplines. Narrative synthesis has become an increasingly accepted technique, across different disciplines, including management, of summarizing and combining both qualitative and empirical information addressing various aspects of the same phenomenon in order to provide a larger picture of that phenomenon (Denyer and Tranfield, 2006). Rumrill and Fitzgerald (2001) identify four key objectives for a narrative synthesis:

- to develop or advance theoretical models;
- to identify, explain, and provide perspectives on complicated or controversial issues;
- to present new perspectives on important and emerging issues; and
- to provide information that can assist practitioners in advancing "best" practice.

Unlike meta-analyses, where there must be a fit between the nature and quality of information sources, narrative synthesis accommodates differences in the questions and designs of various studies, as well as differences in the context of information. Narrative synthesis is especially valuable when information includes both quantitative and qualitative sources (Cassell and Symon, 1994). Moreover, narrative synthesis provides deep and rich information (Light and Pillmer, 1984) that is not obtained by other methods.

Key principle

A comprehensive model of leading sustainability efforts was built by synthesizing both quantitative and qualitative information from various fields.

How the model is presented

Our discussion throughout the chapters of this book follows a top-down flow through the Leading the Sustainable Organization model. The process begins with the articulation of sustainability as part of the mission, values, goals, and strategy of the organization. The human resources (HR) value chain supports this mission with efforts to attract, retain, and engage a talented workforce. The macro-level practices set the tone for creating an engaged workforce, but they are not sufficient to create a sustainable enterprise. These macro-level practices must be complemented by a set of micro-level practices. Leader behavior, job characteristics, and challenging goals are the necessary precursors to employee engagement in sustainability efforts. When employees experience engagement in sustainability endeavors, there will be positive in-role (job specific) and extra-role (above and beyond the job) employee performance, as well as positive overall firm performance. Finally, trust acts as an important enhancer of the relationships between the precursors and outcomes of workforce engagement in sustainability efforts.

In practice, leading sustainability efforts is often messier than the top-down flow presented in this book. Multiple iterations of various components of the model are necessary as organizations and their sustainability efforts evolve. Likewise, management often will find the need to loop back from one component of the model in order to refine or adjust a previous component before moving forward with their sustainability efforts.

Key principle

Multiple iterations of various components of the model are necessary as organizations and their sustainability efforts evolve.

Why the model is important

The Leading the Sustainable Organization model is a comprehensive model that is an important advancement in the sustainability literature, for several reasons. First, no book has yet combined a comprehensive view of leadership with the development, implementation, and assessment of firms' sustainability efforts. Second, the framework transcends industries and the size of companies. Third, because the components of the model have been developed by drawing on research from various disciplines, academic researchers can use the framework to study sustainability efforts from a multi-level and multi-disciplinary perspective. Thus, the model provides a framework to guide research that is focused in the broader context of overall sustainability efforts. Lastly, several studies have identified what needs to occur during sustainability efforts in order to achieve success (Van Velsor, 2009). However, despite observations on what needs to be done around sustainability, many organizations do not seem to know how to go about it. Therefore, throughout our discussion of the components of the Leading the Sustainable Organization model, practical guidelines, along with easy-to-use tools and templates, are provided for executives and managers seeking to apply sustainability best practices within their organizations.

Key principle

Despite observations on what needs to be done around sustainability, many organizations do not seem to know how to go about it.

Sustainability literature lacks leadership

The research on sustainability to date has been fragmented (Turker, 2009). Table 1.1 provides an overview of the various perspectives that existing sustainability research has addressed.

Table 1.1 Various perspectives of empirical inquiry into corporate sustainability

Definitions of sustainability	Bowen, 1953; Davis, 1960; Frederick, 1960; McGuire, 1963; Davis and Blomstrom, 1966; Eells and Walton, 1974; Sethi, 1975; Fitch, 1976; Carroll, 1979, 1999; Jones, 1980; Epstein, 1987; Devinney, 2009; Turker, 2009; Strategic Direction, 2010
Stages and levels of a firm's corporate sustainability endeavors	Zadek, 2004; Munilla and Miles, 2005; Mirvis and Googins, 2006; Castello and Lozano, 2009
Sustainability mission	Castello and Lozano, 2009; Jacopin and Fontrodona, 2009
Sustainability values	Castello and Lozano, 2009; Jacopin and Fontrodona, 2009; Morsing and Oswald, 2009; White, 2009
Sustainability strategy	Porter and Kramer, 2002, 2006; Berns *et al.*, 2009; Castello and Lozano, 2009; Jacopin and Fontrodona, 2009; Morsing and Oswald, 2009; Perera Aldama *et al.*, 2009; Siegel, 2009; Sloan, 2009; Spitzeck, 2009; White, 2009
Sustainability operations	Porter and Kramer, 2006; Lacy *et al.*, 2009; Sloan, 2009; Golicic *et al.*, 2010
Stakeholder involvement	Ahern, 2009; Hind *et al.*, 2009; Lacy *et al.*, 2009; Morsing and Oswald, 2009; Rok, 2009; Sloan, 2009; Turker, 2009
Measurement, performance, and reporting	McGuire *et al.*, 1988; Orlitzky *et al.*, 2003; Orlitzky, 2005; Wu, 2006; Bonini *et al.*, 2009; Carleton, 2009; Lee *et al.*, 2009; Marker *et al.*, 2009; Sloan, 2009

While the number of research studies regarding sustainability has steadily increased since the 1980s, "the literature on sustainability and corporate social responsibility has not paid much attention so far to how leaders enact a corporate sustainability strategy among organizational members" (Morsing and Oswald, 2009: 83). This view is echoed by Van Velsor, who notes that, "While books and articles focusing on CSR abound, there is very little that addresses the leadership aspect, and even less that is based on sound empirical research … Practically speaking, studies are only beginning to surface which identify real leadership practices, systems, and processes that organizations have used to effectively face the challenges posed by moving towards more socially responsible business operations on a global scale" (Van Velsor, 2009: 3). Those who have examined the leadership of corporate sustainability have relied primarily on interviews and case studies (ibid.). These approaches are summarized in Table 1.2. Moreover, the importance of employee engagement in sustainability efforts has received even less attention from researchers than the topic of leading corporate sustainability.

Table 1.2 Various perspectives of empirical inquiry into the leadership of corporate
sustainability

Sustainability leadership competencies	Frankel, 1998; Jackson and Nelson, 2004; Hind *et al.*, 2009; Kakabadse *et al.*, 2009; Lacy *et al.*, 2009
Sustainability leadership tasks	Quinn and Dalton, 2009
Patterns of sustainability leadership practice	Caspary, 2009; D'Amato and Roome, 2009; Hargett and Williams, 2009; Hopkins, 2009; Morgan *et al.*, 2009; Morsing and Oswald, 2009
Sustainability leadership styles	Waldman *et al.*, 2006; Basu and Palazzo, 2008
Sustainability leadership values	Ahern, 2009; Rok, 2009

Key principle

Research and writing about sustainability has not yet focused on how
leaders engage a workforce in corporate sustainability efforts.

The current state of sustainability

Companies appear to be undertaking sustainability initiatives for a wide variety of
reasons. One of the main factors contributing to the fast-growing corporate focus
on sustainability is substantial public interest. Sustainability is shaping everything,
from the foods we eat, to the places we live and work, to the endeavors in which
we engage as individuals and groups. In addition to the public's interest, changes
in legislation, pressure from stakeholders including communities and non-
governmental organizations (NGOs), and concern for their reputation have
compelled even the most reluctant managers to undertake sustainability efforts.
As with the other waves that hit organizations in the latter decades of the previous
century, many initially thought that sustainability was a passing fad; yet it is now
clear that sustainability is here to stay.

Some companies are tying their sustainability initiatives to branding and image
strategies; some see sustainability as a key element in their attempts to differentiate
themselves from competitors; while others see sustainability as an extended cost-
saving exercise. Ford Motor Company was an early mover on the sustainability
front, and in 2000 was one of the first industrial companies to produce a
sustainability report (Bonini and Gorner, 2010). Others, such as Dave Steiner,
CEO of Waste Management, oversaw a strategic planning effort to identify ways
of profiting from sustainability activities. This effort resulted in a greater focus on
its recycling business, on turning waste into energy, and on improving the fuel
efficiency of its trucks. FedEx has also implemented efforts to improve the fuel
efficiency of its planes and vehicles, and to use more alternative fuels. Dow

Chemical executives are overseeing the development of a second set of ten-year sustainability goals, building on goals originally developed in 1994. From 1994 to 2005, Dow invested $1 billion to reduce its energy consumption and improve company-wide water and energy productivity, garnering $4.3 billion in cost savings. Dow's cost savings have continued from these efforts, totaling over $8 billion by the beginning of 2009 (Bonini and Gorner, 2010).

In a recent interview, Stuart Rose, the CEO of Marks and Spencer (M&S), called sustainability as an "absolute must" if the company is to be a successful business in the twenty-first century. Rose believes that a public commitment to sustainability brings new opportunities to connect with customers and employees. M&S also sees sustainability as a differentiation strategy and is proactively seeking to address its customers' most pressing concerns. In 2007, the company implemented "Plan A," a $300 million eco-plan that identified 100 actions the company should take to increase its sustainability. Twenty months into Plan A, the company has seventy-seven of the firm's 100 Plan A actions under way, with fourteen completed (Rose, 2010).

Sustainability initiatives are not limited to cost-cutting and differentiation strategies. Many companies are now integrating sustainability into their value statements. For example, Novo Nordisk, a healthcare company, includes a commitment to environmentally and socially responsible business conduct in all organizational practices as part of its corporate values (Morsing and Oswald, 2009). Wilh. Wilhelmsen, a leading global provider of maritime services, has also incorporated sustainability into the company's values, along with its management training and the firm's balanced scorecard measures (Hargett and Williams, 2009). The retailer Summit Gear Cooperative incorporates sustainability into its business by educating both staff and customers about environmentally friendly choices, and it has a five-year sustainability agenda addressing each area of its supply chain (Carleton, 2009). UnitedHealth uses a social responsibility dashboard that includes metrics for environmental impact, employee–community involvement, and stakeholders' perspectives on social responsibility and community giving (Bonini *et al.*, 2009).

In our survey of 124 employees and managers from seventeen different industries regarding the "Current state of sustainability leadership" (see Appendix A for a detailed discussion of the survey results), we found a spectrum of opinions about the motivations behind executives' pursuit of sustainability (see Figure 1.2).

Most survey participants believe that the number one reason their company's leaders are pursuing sustainability initiatives is "to gain competitive advantage" (35% of respondents). This would suggest that, for these firms, sustainability efforts are driven by market forces. In contrast to a competitive market orientation, one-fifth of respondents feel that their leaders pursue sustainability in order "to manage risk" (20%), demonstrating that the leaders of these firms see sustainability as something they have to do because of regulatory requirements. Beyond market or regulatory motivations, approximately one-quarter of the respondents believe that their company's leaders view sustainability as a moral imperative, with 13% indicating that their leaders see sustainability as "the right thing to do for society",

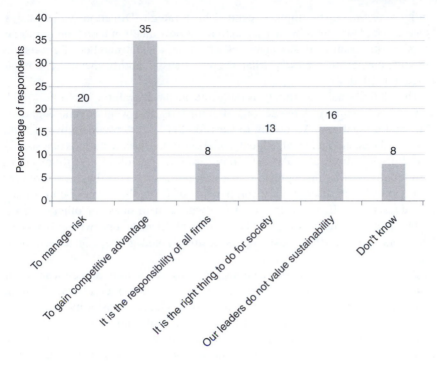

Figure 1.2 Perceptions of both managers and employees as to why their firm pursues
sustainability

and 8% indicating that their leadership see sustainability as "the responsibility of
all firms." Finally, 16% of survey participants believe that their leaders "do not
value sustainability," while 8% feel that they "don't know" the motivations of
their leaders for pursuing sustainability efforts.

Key principle

The various reasons companies pursue sustainability efforts range from
compliance, to cost savings, to brand-building and differentiation.

Various definitions of sustainability

Despite the momentum of sustainability efforts in the corporate sector, corporate
sustainability remains an ill-defined term. As noted above, sustainability, CSR,
"going green," and the "triple bottom line" are all common phrases that are used
to refer to the phenomenon. Devinney (2009) focuses on CSR, and argues that
the term creates confusion not only among firms attempting to implement the
concept, but also for scientific inquiry into the topic. The lack of definitional

clarity surrounding the topic is echoed by Strategic Direction, which states, "One hundred percent of the FTSE companies now mention it on their corporate website. But despite all this increased attention, a great number of executives are still unclear on precisely what the word sustainability means" (Strategic Direction, 2010: 27).

This definitional uncertainty is reflected in the literature. McGuire (1963) viewed responsibility as the actions of companies that went beyond economic and legal obligations. According to Carroll (1979), the responsibility of firms includes the economic, legal, ethical, and discretionary expectations that society has of organizations at a particular point in time. Carroll (1999) followed his earlier definition by distinguishing the economic aspects of business from the legal, ethical, and discretionary components of CSR; the former being what a firm does for itself, while the latter are what a firm does for others. Turker (2009: 189) also defines a socially responsible firm as one whose corporate behaviors "aim to affect stakeholders positively and go beyond its economic interest."

Porter and Kramer (2006) use both CSR and sustainability in their discussion, and offer a definition for each (ibid.: 3). They use a definition of CSR from the US-based, non-profit association Business for Social Responsibility, which asks its members to "achieve commercial success in ways that honor ethical values and respect people, communities, and the natural environment;" for sustainability, they use the World Business Council for Sustainable Development's definition, "meeting the needs of the present without compromising the ability of future generations to meet their own needs."

For consistency, we use the term sustainability throughout this book. However, the Leading the Sustainable Organizations model that the book is structured around is intended to provide a framework for the development, implementation, assessment, and study of all aspects of sustainability and related concepts.

Key principle

Sustainability remains an ill-defined term, and there are several monikers by which the concept is referred to.

Sustainability and workforce engagement are reciprocal

Although our discussion presents a linear, top-down flow through the Leading the Sustainable Organization model, in reality workforce engagement and corporate sustainability are reciprocal and mutually reinforcing. At the macro-level, incorporating sustainability in a firm's mission, strategy, and values, as well as reinforcing these through a well designed HR value chain, sets the stage for engaging a firm's workforce in its sustainability efforts. These macro-level, organization-wide practices help create a corporate reputation that makes a firm attractive to potential employees who are seeking an opportunity to work for an

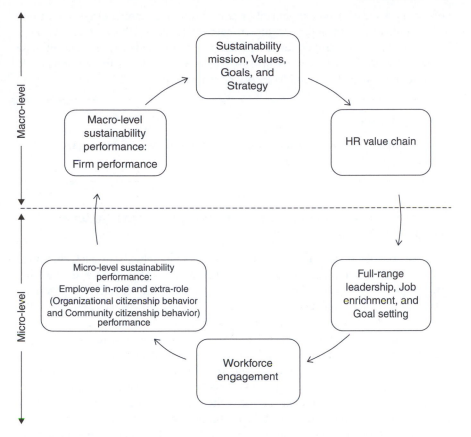

Figure 1.3 Virtuous cycle of sustainability and workforce engagement

organization that is making a positive impact on the communities and environments in which it operates.

But macro-level components alone are not enough to achieve results. Micro-level, manager-to-employee leadership actions are also required. These actions are comprised of full-range leadership, work enrichment, and goal setting. Implementing both the macro- and micro-level components of the model leads to two levels of results. At the micro-level, results are exhibited as employee in-role (job-related) and extra-role (beyond an employee's immediate job) performance; at the macro-/firm level, results are manifested as a company's financial and reputational performance.

The reciprocal nature of corporate sustainability and workforce engagement creates a reinforcing virtuous cycle. The cycle occurs as employees become fully engaged in the firm's sustainability efforts, create results, and help to shape the firm's future sustainability mission, strategy, and value-setting; thus beginning the cycle again with even higher levels of public reputation, sustainability expectations, and performance. Employee participation in firm direction setting creates more ideas, fosters organizational innovation, enhances employee buy-in to firm initiatives, and improves implementation (Hamel, 1996); and all of these have important implications for the development, implementation, and ultimate success of a firm's sustainability efforts. Figure 1.3 illustrates the virtuous cycle of corporate sustainability and workforce engagement.

Key principle

The reciprocal nature of corporate sustainability and workforce engagement creates a reinforcing virtuous cycle.

A macro- and micro-approach to sustainability at the National Geographic Society

Through both macro- and micro-organizational practices, the National Geographic Society is incorporating sustainability throughout the organization and achieving measurable results. At the macro-/organization-wide level, the National Geographic Society launched its "GoGreen" initiative in February 2007, with the stated aim to be "an international leader for global conservation and environmental sustainability." It has developed sustainability initiatives related to water, energy, recycling, and employee programs that are in line with the organization's mission "to inspire people to care about the planet." As part of its HR value chain to support sustainability, management has installed an employee education process to "encourage green behaviors at the office and at home," and the Society encourages alternative commuter transportation opportunities and flexible work options, including discounted parking for hybrid vehicles and car-poolers, pre-tax incentives for public transit or certain qualifying van-pools, and flexible work options such as telecommuting, compressed work weeks, and flexi-hours (National Geographic Society, 2011).

At the micro-level, all employees of the National Geographic Society and *National Geographic* channel were invited to help the organization "walk the talk." The Society established seven cross-divisional subcommittees to help the organization pursue a "path of sustainability, a path we are all committed to and that stands as a shining example of employee-driven change" (ibid.). The organization's sustainability accomplishments since the beginning of its 2007 effort are identified in Table 1.3.

Beyond the accomplishments identified in Table 1.3, because the National Geographic Society has "engaged its staff to be participants in finding sustainable solutions for the Society's operations," the organization has achieved measurable sustainability improvements, including:

- using 2.5 million fewer kilowatt hours of electricity each year;
- saving 4.7 million gallons of water each year; and
- expanding recycling of all materials to 66% of all products used.

Table 1.3 Various National Geographic Society sustainability accomplishments

2007	– Joined the US EPA Climate Leaders
	– Measured the carbon footprint of the Society's headquarters
	– Certified one of our headquarters buildings as Energy Star (US EPA) compliant
	– Initiated two-sided copying
	– Introduced local organic foods and eliminated plastics from our cafeteria
	– Began composting all food waste and non-recyclable paper products, including bathroom hand towels
	– Installed video conferencing and reduced staff business travel by 20 percent
	– Carbon footprinted all of our Expedition travel (educational travel service)
	– Implemented an alternative commuting plan to encourage car-pools, mass transit use, work from home, etc.
	– Committed to developing a comprehensive carbon assessment of all our products
2008	– Developed a carbon footprint for all the locations in North America where the Society does business
	– Calculated the carbon footprint for all Society air travel, accommodations, and rental car use
	– Implemented Green Fridays that close the office ten days in summer to reduce energy use in our buildings
	– Announced a goal to reduce business travel by an additional 20 percent
	– Purchased offsets for all Expeditions travel, and built that into the program cost
	– Implemented after-hours computer hibernation that shut off all related equipment
	– Eliminated all plastic clamshells from cafeteria and all bottled water from our campus
	– De-bulbed all buildings of excessive lighting and installed CFLs
	– Bought wind-power renewable energy credits to offset all electrical use in our buildings (signed a five-year contract)
	– Eliminated unsustainably harvested seafood (shrimp, tuna, salmon) from our cafeteria
2009	– Achieved carbon neutrality for our buildings
	– Joined World Wildlife Fund's Climate Saver Program (made a commitment to reduce our electrical use by 10 percent)
	– Achieved LEED EB Gold Certification for our buildings (in 2003 we were the first LEED EB silver certified building in the US)
	– Qualified our entire campus to be Energy Star certified
	– Completed a very comprehensive life cycle assessment (LCA) on *National Geographic* magazine
	– Calculated our energy reduction that showed the results of a ten-year initiative. Found electricity consumption was cut by 18 percent; water usage was down by 20 percent; gas usage was cut by 3 percent
	– Calculated the carbon footprint for all Society products and services, including TV and Channels programs
	– Held training classes for safe bike commuting, installed additional bike racks and promoted Bike-to-Work Day, which culminated in the Society being designated a Bike Friendly Business
	– Expanded local and organic offerings in the cafeteria
	– Changed the temperature setting of our buildings (down in winter, up in summer)

2010   – Upgraded our Energy Management System allowing building services and
          engineering staff to monitor and better control energy use in our facilities
       – Launched EnerNoc energy reduction initiative – a tangible effort that engaged
          staff at all levels in energy reduction efforts
       – Implemented/updated 20 LEED operating policies for the complex
       – Achieved 60 percent recycling rate
       – Developed sustainability policies for travel, general purchasing, and paper
          purchasing
       – Expanded composting efforts by placing compost bins on every floor
       – Provided and promoted locally sourced and organic food options in the
          headquarters cafeteria
       – Launched the Society's *Sustainability* public-facing website
       – Calculated the Society's total carbon emission for all products and services
       – Implemented a commuter strategy program that includes:
              — mass transit commuter subsidy
              — parking discounts for LEED certified green cars
              — increased discounts for car-pools

Source: www.nationalgeographic.com

Moreover, the management feels that the organization has "picked many of the
low-hanging fruit and now must reach higher to achieve the reduction in energy
use we have targeted" (ibid.). To do that, their 2011 sustainability goals include:

•   continue to develop an energy-reduction plan and strategy;
•   continue efforts to increase local and organic food sources for the headquarters'
    cafeteria;
•   develop more offset programs for products and services;
•   investigate possibilities of an on-site biogas compodigester or fuel cell;
•   organize a Society-wide volunteer day to help "green" a local school;
•   increase the recycling rate to 70%.

Key principle

Through both macro- and micro-organizational practices, the National
Geographic Society is incorporating sustainability throughout the
organization and achieving measurable results.

Summary

•   The path to successful sustainability efforts hinges on both macro (organization-
    wide) and micro (manager-to-employee) organizational factors.
•   There is a clear need for a leadership model that facilitates employee
    engagement in a firm's sustainability efforts.
•   The best way to build a comprehensive leadership model for such a complex
    undertaking as corporate sustainability is by combining both quantitative and
    qualitative information.

- Multiple iterations of various components of the model are necessary as organizations and their sustainability efforts evolve.
- Despite observations on what needs to be done around sustainability, many organizations do not seem to know how to go about it.
- Research and writing about sustainability have not yet focused on how leaders engage a workforce in corporate sustainability efforts.
- The various reasons why companies pursue sustainability efforts range from compliance, to cost savings, to brand building and differentiation.
- Sustainability remains ill-defined, and there are several terms for the concept, including: sustainability, corporate social responsibility (CSR), "going green", and the "triple bottom line."
- The reciprocal nature of corporate sustainability and workforce engagement creates a reinforcing virtuous cycle.
- The National Geographic Society offers a good example of an organization pursuing sustainability at both the macro (organization-wide) and micro (manager-to-employee) levels.

Discussion questions

- How do you define sustainability?
- What is the current emphasis on sustainability in your industry?
- How has sustainability evolved in your industry?
- What is the general history of sustainability within your organization?
- How has your organization gone about implementing sustainability initiatives? What was the role of managers (both top and mid-level) in this process? What is the role of employees in the process?
- How does your organization measure the outcomes of your sustainability initiatives? Do you rely strictly upon internal metrics? Are there external measures, such as certifications and endorsements, which your organization considers to be relevant outcome measures?
- What challenges has your company experienced in implementing its sustainability efforts?
- What steps did your organization take to overcome those challenges?
- Why is it important to reinforce sustainability through both macro (firm-wide) and micro (manager-to-employee) levels?

Key tool

Leading the Sustainable Organization – rapid assessment

Completing the following scorecard will provide a quick, high-level view of the degree to which the key components of the Leading the Sustainable Organization model are being utilized to engage the workforce in a firm's sustainability efforts.

Table 1.4 Leading the Sustainable Organization Rapid Assessment

Component	Rating (0 = poor, 10 = excellent)	Notes/ rationale
1 Sustainability is included in the company's mission, values, goals, and strategy		
2 The company's leaders are committed to sustainability		
3 Sustainability is reinforced at the macro-/ firm-wide level (through the HR processes of the firm)		
4 Sustainability is reinforced at the micro-/ manager to employee level (through leader behavior, job characteristics, and challenging goals)		
5 Employees trust that the firm's management is committed to sustainability		
6 Employees incorporate sustainability into their day-to-day work		
7 Employees pursue sustainability beyond their day-to-day work		
8 The company incorporates sustainability into the way it operates		
9 The company performs better financially because of incorporating sustainability into the way it operates		
10 The company has been recognized by external sources for excellence in its sustainability efforts		
Total score		

Steps to complete the assessment

1. Rate each item on a scale of 0 (poor) to 10 (excellent).
2. Make notes for each item to explain the rationale for the numerical rating.
3. Add all ten scores to obtain a total score (maximum = 100).

Rating scale

- 0–20 = poor (significant improvement needed across most or all components)
- 21–40 = below average (improvement needed in several components)
- 41–60 = average (identify areas of weakness and adjust)
- 61–80 = above average (identify areas that can still be improved)
- 81–100 = excellent (continuously review and refine each component as the firm's sustainability efforts evolve)

References

Ahern, G. (2009) "Implementing environmental sustainability in ten multinationals," *Corporate Finance Review*, 13(6): 27–32.

Basu, K. and Palazzo, G. (2008) "Corporate social responsibility: a process model of sense-making," *Academy of Management Review*, 33(1): 122–136.

Berns, M., Townend, A., Khayat, Z., Balagopal, B., Reeves, M., Hopkins, M.S., and Kruschwitz, N. (2009) "Sustainability and competitive advantage," *MIT Sloan Management Review*, 51(1): 19–26.

Bonini, S. and Gorner, S. (2010) "Making sustainability real," *McKinsey Quarterly*, 1: 92–93.

Bonini, S., Koller, T.M., and Mirvis, P.H. (2009) "Valuing social responsibility programs," *McKinsey Quarterly*, 4: 65–73.

Bowen, H.R. (1953) *Social Responsibilities of the Businessman*, Harper & Row, New York.

Carleton, K.L. (2009) "Framing sustainable performance with the six-p," *Performance Improvement*, 48(8): 37–44.

Carroll, A.B. (1979), 'A three dimensional conceptual model of corporate social performance," *Academy of Management Review*, 4(4): 497–505.

——(1999) "Corporate social responsibility: evolution of a definitional construct," *Business & Society*, 38(3): 268–295.

Caspary, G. (2009) "Improving sustainability in the financing of large infrastructure projects: What role for leaders?," *Corporate Governance*, 9(1): 58–72.

Cassell, C. and Symon, G. (1994) *Qualitative Methods in Organizational Research: A Practical Guide*, Sage, London.

Castello, I. and Lozano, J. (2009) "From risk management to citizenship corporate social responsibility: analysis of strategic drivers of change," *Corporate Governance*, 9(4): 373–385.

Cook, D.J., Mulrow, C.D., and Haynes, R.B. (1997) "Systematic reviews: synthesis of best evidence for clinical decisions," *Annals of Internal Medicine*, 126(5): 376–380.

D'Amato, A. and Roome, N. (2009) "Toward an integrated model of leadership for corporate responsibility and sustainable development: a process model of corporate responsibility beyond management innovation," *Corporate Governance*, 9(4): 421–434.

Davis, K. (1960) "Can business afford to ignore social responsibilities?," *California Management Review*, 2(3): 70–76.

Davis, K. and Blomstrom, R.L. (1966) *Business and its Environment*, McGraw-Hill, New York.

Denyer, D. and Tranfield, D. (2006) "Using qualitative research synthesis to build an actionable knowledge base," *Management Decision*, 44(2): 213–227.

Devinney, T.M. (2009) "Is the socially responsible corporation a myth? The good, the bad, and the ugly of corporate social responsibility," *Academy of Management Perspectives*, 23(2): 44–56.

Edwards, A. (2005) *The Sustainability Revolution: Portrait of a Paradigm Shift*, New Society Publishing, Gabriola Island, BC, Canada.

Eells, R. and Walton, C. (1974) *Conceptual Foundations of Business*, 3rd edn, Irwin, Burr Ridge, IL.

Elkington, J. (1998) *Cannibals with Forks: The Triple Bottom Line of 21st Century Business*, New Society Publishing, Gabriola Island, BC, Canada.

Epstein, E.M. (1987) "The corporate social policy process: beyond business ethics, corporate social responsibility, and corporate social responsiveness," *California Management Review*, 29(3): 99–114.

Fitch, H. G. (1976) "Achieving corporate social responsibility," *Academy of Management Review*, 1(1): 38–46.

Frederick, W. C. (1960) "The growing concern over business responsibility," *California Management Review*, 2(4): 54–61.

Frankel, C. (1998) *In Earth's Company: Business, Environment and the Challenge of Sustainability*, New Society Publishing, Gabriola Island, BC, Canada.

Golicic, S.L., Boerstler, C.N., and Ellram, L.M. (2010) "'Greening' transportation in the supply chain," *MIT Sloan Management Review*, 51(2): 46–55.

Hamel, G. (1996) "Strategy as revolution," *Harvard Business Review*, 74(4): 69–82.

Hammersley, M. (2001) "On 'systematic' reviews of research literatures: a 'narrative' response to Evans and Benefield," *British Educational Research Journal*, 27(5): 543–554.

Hargett, T.R. and Williams, M.F. (2009) "Wilh. Wilhelmsen Shipping Company: moving from CSR tradition to CSR leadership," *Corporate Governance*, 9(1): 73–82.

Hind, P., Wilson, A. and Lenssen, G. (2009) "Developing leaders for sustainable business," *Corporate Governance*, 9(1): 7–20.

Hopkins, M. (2009) "8 reasons sustainability will change management (that you never thought of)," *MIT Sloan Management Review*, 51(1): 27–30.

Jackson, I.A. and Nelson, J. (2004) *Profits with Principles: Seven Strategies for Delivering Value with Values*, Currency, New York.

Jacopin, T. and Fontrodona, J. (2009) "Questioning the corporate responsibility (CR) department alignment with the business model of the company," *Corporate Governance*, 9(4): 528–536.

Jones, T.M. (1980) "Corporate social responsibility revisited, redefined," *California Management Review*, 22(3): 59–67.

Kakabadse, N.K., Kakabadse, A.P., and Lee-Davies, L. (2009) "CSR leaders road-map," *Corporate Governance*, 9(1): 50–57.

Lacy, P., Arnott, J., and Lowitt, E. (2009) "The challenge of integrating sustainability into talent and organization strategies: investing in the knowledge, skills and attitudes to achieve high performance," *Corporate Governance*, 9(4): 484–494.

Lee, D.D., Faff, R.W., and Langfield-Smith, K. (2009) "Revisiting the vexing question: Does superior corporate social performance lead to improved financial performance?," *Australian Journal of Management*, 34(1): 21–49.

Light, R.J. and Pillmer, D.B. (1984) *Summing Up*, Harvard Business School Press, Boston, MA.

Marker, A., Johnsen, E., and Caswell, C. (2009) "A planning and evaluation six-pack for sustainable organizations: the six-p framework," *Performance Improvement*, 48(8): 27–34.

McGuire, J.W. (1963) *Business & Society*, McGraw-Hill, New York.

McGuire, J.B., Sundgren, A., and Schneeweis, T. (1988) "Corporate social responsibility and firm financial performance," *Academy of Management Journal*, 31(4): 854–872.

Mirvis, P. and Googins, B. (2006) "Stages of corporate citizenship," *California Management Review*, 48(2): 104–126.

Morgan, G., Ryu, K., and Mirvis, P. (2009) "Leading corporate citizenship: governance, structure, systems," *Corporate Governance*, 9(1): 39–49.

Morsing, M. and Oswald, D. (2009) "Sustainable leadership: management control systems and organizational culture in Novo Nordisk A/S," *Corporate Governance*, 9(1): 83–99.

Munilla, L. and Miles, M.P. (2005) "The corporate social responsibility continuum as a component of stakeholder theory," *Business and Society Review*, 110(4): 371–387.

National Geographic Society (2011) "Sustainability at National Geographic," http://environment.nationalgeographic.com/environment/national-geographic-sustainability

Orlitzky, M. (2005) "Payoffs to social and environmental performance," *Journal of Investing*, 14(3): 48–51.

Orlitzky, M., Schmidt, F.L., and Rynes, S.L. (2003) "Corporate social and financial performance: a meta-analysis," *Organization Studies*, 24(3): 403–441.

Perera Aldama, L.R., Amar, P.A., and Trostianki, D.W. (2009) "Embedding corporate responsibility through effective organizational structures," *Corporate Governance*, 9(4): 506–516.

Porter, M.E. and Kramer, M.R. (2002) "The competitive advantage of corporate philanthropy," *Harvard Business Review*, 80(12): 56–65.

——(2006) "Strategy and society: the link between competitive advantage and corporate social responsibility," *Harvard Business Review*, 84(12): 78–92.

Quinn, L. and Dalton, M. (2009) "Leading for sustainability: Implementing the tasks of leadership," *Corporate Governance*, 9(1): 21–38.

Rok, B. (2009) "Ethical context of the participative leadership model: taking people into account," *Corporate Governance*, 9(4): 461–472.

Rose, S. (2010) "Business and sustainability," *The British Journal of Administrative Management*, Winter: 16–17.

Rumrill, P.D. and Fitzgerald, S.M. (2001) "Using narrative reviews to build a scientific knowledge base," *Work*, 16: 165–170.

Savitz, A.W. and Weber, K. (2006) *The Triple Bottom Line: How Today's Best-Run Companies Are Achieving Economic, Social and Environmental Success – And How You Can Too*, Jossey-Bass, San Francisco, CA.

Sethi, S.P. (1975) "Dimensions of corporate social performance: an analytic framework," *California Management Review*, 17(3): 258–264.

Siegel, D.S. (2009) "Green management matters only if it yields more green: an economic/strategic perspective," *Academy of Management Perspectives*, 23(3): 5–16.

Sloan, P. (2009) "Redefining stakeholder engagement: from control to collaboration," *Journal of Corporate Citizenship*, 36: 25–40.

Spitzeck, H. (2009) "The development of governance structures for corporate responsibility," *Corporate Governance*, 9(4): 495–505.

Strategic Direction (2010) "What does sustainability mean? Debate, innovation and advice around a key but complicated concept," *Strategic Direction*, 6(2): 27–30.

Turker, D. (2009) "How corporate social responsibility influences organizational commitment," *Journal of Business Ethics*, 89: 189–204.

Van Velsor, E. (2009) "Introduction: leadership and corporate social responsibility," *Corporate Governance*, 9(1): 3–6.

Waldman, D.A., Siegel, D.S., and Javidan, M. (2006) "Components of CEO transformational leadership and corporate social responsibility," *Journal of Management Studies*, 43: 703–1725.

White, P. (2009) "Building a sustainability strategy into the business," *Corporate Governance*, 9(4): 386–394.

Wu, M. (2006) "Corporate social performance, corporate financial performance, and firm size: a meta-analysis," *Journal of American Academy of Business*, 8(1): 163–171.

Zadek, S. (2004) "The path to corporate responsibility," *Harvard Business Review*, 82(12): 125–132.

Part I
Development

2 Direction setting

Mission, values, goals and strategy

The process of leading the sustainable organization begins at the macro-/ organization-wide level with the articulation of sustainability as part of the firm's mission, values, goals, and strategy (see Figure 2.1). This level is where the process of incorporating sustainability into the organization's **strategic management** practices begins.

Strategic management is not just the organization's strategic plan. Rather, it is the total sum of a firm's plans, goals, and actions, leading to measurable results. This more integrated view of the firm is in contrast to traditional strategic planning, which is characterized by the systematic formulation of strategies geared toward the achievement of organizational goals (Mintzberg, 1987). Research has shown that firms which put strategic management into practice typically outperform those organizations that do not (Andersen, 2000; Miller and Cardinal, 1994; Pekar and Abraham, 1995). In a survey of more than fifty organizations, Wilson (1994) found that the top three benefits of strategic management are:

- a clearer direction for the company;
- a sharper focus on what is strategically important; and
- improved understanding of a rapidly changing environment.

Key principle

Starting with the firm's mission, values, goals, and strategy begins the process of incorporating sustainability throughout the organization.

A systemic perspective

Strategic management entails taking a systemic view of both the internal organization and how the firm interacts with the external environment. This systemic view helps executives set, implement, and measure the future direction of the firm. Strategic management includes strategy formulation (strategic planning), and it also involves strategy implementation and evaluation. Moreover, to ensure that management and employees are all moving in the same direction,

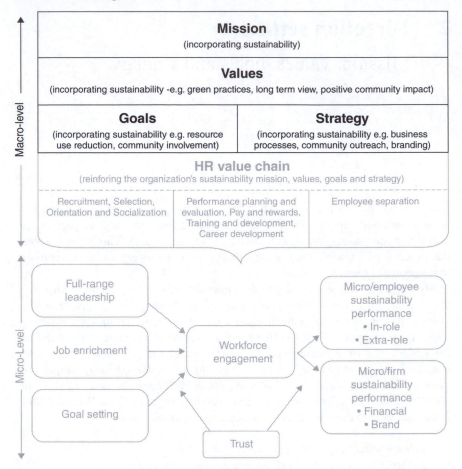

Figure 2.1 Begin leading the sustainable organization by including sustainability as part of the firm's mission, values, goals, and strategy

each of the firm's internal strategic management components must be in alignment with one another. Figure 2.2 illustrates the relationship and interaction between a firm's internal and external environments, and the alignment required between each component of the firm's overall strategic management approach.

The process of strategic management can be either formal or informal. Larger companies typically pursue a more formal process. For example, a survey of executives at 708 companies across five continents found that the two most popular management tools are strategic planning (89% of firms surveyed) and developing mission statements (84% of firms surveyed) (Rigby, 2003). Although smaller companies often pursue a more informal approach to strategic management (Wheelen and Hunger, 2008), small and medium-sized firms with higher levels of formal planning have been found to achieve better financial performance than those with lower levels of formal planning (Rue and Ibrahim, 1998). Moreover,

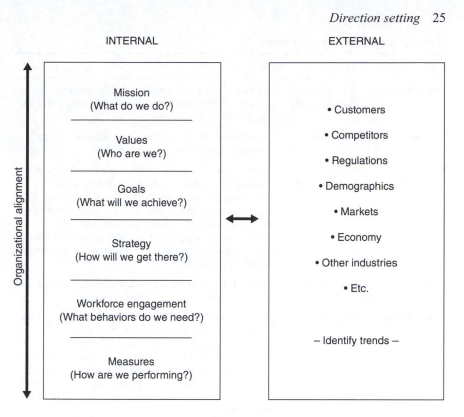

INTERNAL   EXTERNAL

Organizational alignment

Mission
(What do we do?)

Values
(Who are we?)

Goals
(What will we achieve?)

Strategy
(How will we get there?)

Workforce engagement
(What behaviors do we need?)

Measures
(How are we performing?)

• Customers

• Competitors

• Regulations

• Demographics

• Markets

• Economy

• Other industries

• Etc.

– Identify trends –

Figure 2.2 Systemic perspective of strategic management

because of the systemic nature of strategic management (turning strategic plans into measurable results through effective implementation), to be effective a firm's strategic management approach (whether planning is done formally or informally) should include front-line supervisors and employees (Hamel, 1996). Without the involvement and commitment of employees at all levels of the firm, even the best strategic plans will fall short of their intended goals.

Key principle

To ensure management and employees are all moving in the same direction, each of the firm's internal strategic management components must be in alignment with one another.

Mission – what we do

In general, a mission identifies how a firm defines itself, and establishes the priorities of the organization (Jacopin and Fontrodona, 2009). More specifically, in terms of its role in sustainability, a mission identifies the self-assigned role of

Table 2.1 Example mission statements

Company	Industry	Mission
Dell Computers	Electronics manufacturing	To be the most successful computer company in the world at delivering the best customer experience in markets we serve.
Google	Technology	To organize the world's information and make it universally accessible and useful.
Sony	Electronics manufacturing	Sony is committed to developing a wide range of innovative products and multimedia services that challenge the way consumers access and enjoy digital entertainment. By ensuring synergy between businesses within the organization, Sony is constantly striving to create exciting new worlds of entertainment that can be experienced on a variety of different products.
Starbucks	Food service	To inspire and nurture the human spirit – one person, one cup and one neighborhood at a time.
Exxon Mobil	Oil and gas	Exxon Mobil Corporation is committed to being the world's premier petroleum and petrochemical company. To that end, we must continuously achieve superior financial and operating results while adhering to the highest standards of business conduct. These unwavering expectations provide the foundation for our commitments to those with whom we interact.
Walmart	Retail	We save people money so they can live better.

Source: Company websites

the organization in relation to the society in which it operates (Castello and Lozano, 2009). As illustrated in Figure 2.2, a mission answers the question "What do we do as an organization?" In their book *Winning*, Welch and Welch (2005) provide several characteristics of effective mission statements.

- Effective mission statements balance the possible and the impossible.
- Setting the mission is top management's responsibility.
- Too frequently, mission statements are more hot air than real action.
- Organizations don't reach their full potential if the mission is just a platitude on the wall.

An organization's mission communicates what the firm provides. A well designed mission statement defines the company's primary, distinctive purpose, setting the firm apart from other similar organizations. Moreover, a mission statement promotes a set of shared expectations among employees and communicates a public persona to external stakeholders such as the community, consumers, and investors (Wheelen and Hunger, 2008). Too often, a mission statement is long, drawn out, and all-inclusive, illustrating management's attempt to be all things to all people. Such missions do nothing more than confuse those who even take the time to read them. Mission statements should be short, understandable by the entire organization, and repeatable. Table 2.1 lists the mission statements of a number of high-profile companies. Beyond the example statements in Table 2.1, an example of a clear, concise, repeatable, and enduring mission statement comes from Newport News Shipbuilding, whose mission has not changed since the company's founding in 1886:

> We shall build good ships here – at a profit if we can – at a loss if we must – but always good ships.

Sustainability and organizational mission

The exclusion or inclusion of sustainability in a firm's mission is indicative of the firm's commitment to the pursuit of sustainability. In our "Current state of sustainability leadership" survey (see Appendix A), we found that 58% of respondents agreed or strongly agreed with the statement "Sustainability is included in my company's mission...", whereas 42% of respondents disagreed, strongly disagreed, or were not sure (see Figure 2.3). These results indicate that many firms' leaders are beginning to view sustainability as being at least part of their company's purpose.

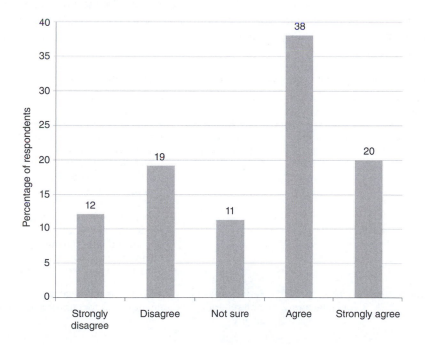

Figure 2.3 Sustainability is included in my company's mission

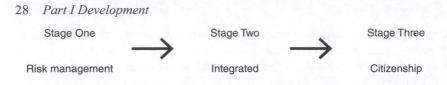

Figure 2.4 Stages of sustainability

Moreover, inclusion of sustainability in a firm's mission can help identify the stage of sustainability development a firm is currently in. Figure 2.4 identifies the three stages of sustainability at which firms can find themselves.

The first stage of commitment to sustainability, is the **risk-management** perspective characterized by intermittent sustainability activity and undeveloped sustainability programs. At the risk-management stage, sustainability is viewed only as a tool to protect a firm's reputation, and sustainability is not included in the firm's mission (Castello and Lozano, 2009).

The second stage of sustainability development is the **integrated** stage, typified by the positioning of sustainability in a firm's operating model. At this stage, firms begin to change their business model to include social and environmental responsibilities. This stage relates to what others call the strategic perspective (Munilla and Miles, 2005), the managerial and strategic stage (Zadek, 2004), or the engaged stage (Mirvis and Googins, 2006) of sustainability. In the integrated stage, the firm actively reflects on ways in which it can use social issue management to gain competitive advantage. There is also inclusion of sustainability within the firm's communications, but not yet in its mission (Castello and Lozano, 2009). Table 2.2 provides examples of companies that, based on espousing separate corporate and sustainability missions, would be classified in the second stage of corporate sustainability.

Table 2.2 Example company mission statements and sustainability mission statements

Company	*Industry*	*Mission*	*Sustainability mission*
Fairmont Hotels	Hospitality	Turning moments into memories for our guests.	Fairmont Hotels & Resorts is committed to environmental protection and sustainability guided by our very own Green Partnership Program. The Partnership, a company-wide stewardship program, strives to minimize our properties' operational impact on the environment through resource conservation and best practices.
Bridgestone	Tire manufacturing	Serving society with superior quality.	To help ensure a healthy environment for current and future generations ...

Company	Industry	Mission	Sustainability mission
Four Seasons Hotels	Hospitality	To be recognized as the company that manages the finest hotels, resorts and residence clubs wherever we locate.	Four Seasons involves employees and guests in the common goal of preserving and protecting the planet. We engage in sustainable practices that conserve natural resources and reduce environmental impact. As importantly, sustainable tourism will enhance and protect the destinations where Four Seasons operates for generations to come.
Nike	Footwear and apparel	To bring inspiration and innovation to every athlete in the world.	Reduce our consumer and our businesses' energy footprint to enable both to thrive in tomorrow's low-carbon economy.
Bristol-Myers Squibb	Bio-pharmaceuticals	To discover, develop and deliver innovative medicines that help patients prevail over serious diseases.	Conducting our business to help patients prevail over serious diseases in a manner that contributes to economic growth, social responsibility and a healthy environment now and in the future.

Source: Company websites

The third stage of corporate sustainability is the **citizenship** stage (ibid.). Others have identified this stage as the civil corporation (Zadek, 2004) or the transforming stage (Mirvis and Googins, 2006). The citizenship stage is characterized by senior leadership's openness to integrating social issues as part of the firm's responsibilities. During this phase, there is a transformation of the firm's business model to assume a role in leading social issues (Logsdon and Wood, 2002; Mirvis and Googins, 2006). The transition to the citizenship stage is driven by the refinement of the company's mission to include sustainability (Castello and Lozano, 2009). Table 2.3 provides examples of company mission statements that incorporate sustainability, indicating that these companies are in the third stage of sustainability.

Key principle

The exclusion or inclusion of sustainability in a firm's mission helps identify the stage of sustainability development the firm is at.

Table 2.3 Example company mission statements that include sustainability

Company	Industry	Mission including sustainability
Ameren	Electrical generation and transmission	To meet our customers' energy needs in a safe, reliable, efficient and environmentally responsible manner by increasing diverse supplier participation in Ameren procurement opportunities.
Patagonia	Apparel	Build the best product, cause no unnecessary harm, use business to inspire and implement solutions to the environmental crisis.
Xanterra Parks and Resorts	Hospitality	To be the industry leader in park and resort hospitality. We are committed to practicing integrity and quality, maintaining positive relationships with our employees and clients, leading in environmental stewardship and creating unforgettable memories for our guests.
Ecolab	Cleaning, sanitizing, food safety and infection control products and services	To be the leading global innovator, developer and provider of cleaning, sanitation and maintenance products, systems, and services. As a team, we will achieve aggressive growth and fair return for our shareholders. We will accomplish this by exceeding the expectations of our customers while conserving resources and preserving the quality of the environment.
New Leaf Paper	Paper manufacturing	To be the leading national source for environmentally responsible, economically sound paper. We supply paper with the greatest environmental benefit while meeting the business needs of our customers. Our goal is to inspire – through our success – a fundamental shift toward environmental responsibility in the paper industry.
Dole Food Company	Food production	Dole Food Company, Inc. is committed to supplying the consumer and our customers with the finest, high-quality products and to leading the industry in nutrition research and education. Dole supports these goals with a corporate philosophy of adhering to the highest ethical conduct in all its business dealings, treatment of its employees, and social and environmental policies.
Whole Foods	Grocery retail	Whole Foods – Whole People – Whole Planet.
Green Mountain Energy	Retail electricity	Change the way power is made.

Source: Company websites

Values – who we are

Organizational values refer to beliefs about the types of goals firm members should pursue, as well as ideas regarding standards of behavior organizational members should use to achieve these goals (Schein, 1993). Values are the basis for the development of organizational norms and expectations that define appropriate behavior by employees in particular situations. Shared values can also provide a source of motivation, commitment, and loyalty among an organization's members (Morsing and Oswald, 2009). Values are at the core of who people are, influencing the choices individuals make, the people they trust, the appeals to which they respond, and how their time and energy are invested (Posner, 2010). Clearly articulated organizational values can help identify the fit between employees and the firm. For example, someone who values teamwork will fit much better in a firm that espouses teamwork as a core value. Numerous studies have found that when an employee's values fit the organization's values, the employee will stay longer and be more productive (Kristof-Brown *et al.*, 2005). Figure 2.2 shows that a firm's values answer the question, "Who are we as an organization?" Example organizational core values include:

- balance – maintaining a positive work and life balance for our workforce;
- diversity – respecting the differences between individuals across our organization;
- teamwork – working together to solve problems and achieve organizational goals;
- fun – celebrating successes, large and small;
- innovation – welcoming new and creative ideas;
- integrity – acting honestly in all we do, without compromising the truth;
- passion – putting our hearts and minds in our work to achieve the best results possible;
- risk taking – encouraging each other to take risks without fear of retribution for failure;
- safety – ensuring a safe and healthy environment for our employees and customers; and
- continuous learning – understanding and applying key lessons gained from our successes and our failures.

Shared values have also been found to be a key component of aligning employees with a firm's sustainability efforts (Hargett and Williams, 2009; Morsing and Oswald, 2009). In their study of management control systems pertaining to sustainability, Morsing and Oswald (2009: 85) assert that company values are the only way to ensure managers do the right thing in all situations around the world. They state that "If all employees share a common understanding of the organization's values and are well trained in what it means to apply those values, they will not have to look to formal policies nor will they be engaging in guess-work to decide how to respond to novel and/or "sticky" problems." Embedding

Table 2.4 Examples of companies' sustainability values

Company	Industry	Sustainability values
Whole Foods	Grocery retail	Caring about our communities and our environment.
P&G	Consumer products	We are accountable for all of our own actions: these include safety, protecting the environment, and supporting our communities.
Nokia	Electronics	Very human: This applies to what we offer customers, how we do business and the impact of our actions and behavior on people and the environment. It is about being very human in the world – making things simple, respecting and caring, even in tough business situations.
Siemens	Electronics and electrical engineering	Responsible: Committed to ethical and responsible actions.
Henkel	Laundry and home care, cosmetics/ toiletries, and adhesive technologies	We are committed to leadership in sustainability: We provide products, technologies and processes that meet the highest standards. We are committed to the safety and health of our employees, the protection of the environment and the quality of life in the communities in which we operate.
Green Mountain Energy	Retail electricity	Sustainability: We are dedicated to the environment and maintaining lasting, mutually beneficial relationships in all aspects of our business.

Source: Company websites

sustainability into a firm's core values helps align employees with the organization's sustainability agenda (Hargett and Williams 2009). Likewise, Rok (2009) found that a firm's values are a vital component in determining the total sustainability motivation held by each employee. Table 2.4 lists examples of sustainability values and the companies that espouse them.

In our survey (Appendix A), we found that 59% of respondents agreed or strongly agreed with the statement "Sustainability is included in my company's values…," whereas 42% of respondents disagreed, strongly disagreed, or were not sure (see Figure 2.5). This signifies that many companies' leaders view sustainability as an important part of their organizational priorities.

Key principle

Embedding sustainability into a firm's core values helps communicate the management and employee behaviors required to put sustainability into practice.

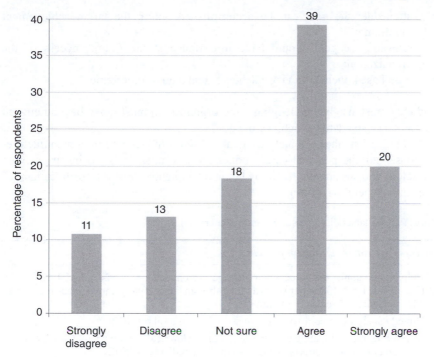

Figure 2.5 Sustainability is included in my company's values
(totals more than 100% due to rounding)

Goals – what we will achieve

Organizational goal setting is considered to be a crucial initial step in the strategic management of a firm. Goal setting provides the foundation for developing a roadmap of organizational activity (the company's strategy), as well as the basis for establishing the metrics that will be used to measure progress (Ransom and Lober, 1999). Etzioni (1960) defines a corporate goal as a desired state of affairs which the organization attempts to realize, or that future state of affairs which the organization as a collectivity is trying to bring about. Likewise, Thompson (1967) states that goals are intended future domains for the organization. Commonly stated organizational goals include profitability, sales and earnings, growth, market share, innovation, customer satisfaction, and employee productivity (Doyle, 1994). Goals are used to communicate the priorities of the firm to all stakeholders. As identified in Figure 2.2, a firm's goals answer the question "What will the organization achieve?" The characteristics of sound organizational goal setting are:

- specific: identifies explicit performance standards (e.g. an increase or decrease of $x\%$);
- measurable: clear metrics can be applied;

- attainable: the goal can be accomplished within the target achievement timeframe;
- relevant: the goal should have importance to the future success of the organization;
- time-based: includes a target achievement date or timeframe;

and they must also be challenging – the organization must move beyond current performance standards to achieve the goal.

Table 2.5 lists the sustainability goals of Seventh Generation, a manufacturer of household and personal care products, whose mission is "to inspire a more conscious and sustainable world by being an authentic force for positive change" (Seventh Generation, 2011).

Table 2.5 Sustainability goals at Seventh Generation

Seventh Generation sustainability goals

- Helping our employees reduce their average personal energy use 20% by 2010.
  (Data totals for 2010 are not yet available, though the company knows it has reached a 14% reduction at the end of Q1 in 2010.)
- Directly sourcing a supply of 100% sustainable palm oil by 2012.
- Identifying and eliminating all persistent and/or chronically toxic chemicals used in the manufacture of our products or found in their ingredients by 2012.
- Reducing our virgin plastic use by 80% by 2014.
- Full Forest Stewardship Council certification of all our virgin pulp by 2015.
  (The wood for almost all of our virgin pulp is currently FSC certified but we have not certified the chain of custody.)
- Making all of our products from 100% renewable plant and mineral sources and ensuring that they are backyard-compostable and/or biodegradable in the marine environment by 2015.
- Reducing our overall greenhouse gas emissions (GHG) 80% by 2050 from a 2005 baseline.
- Reducing our products' life-cycle GHG 15% by 2015 from a 2007 baseline.
  (In 2009, we engineered a 32% drop in GHG emissions per case of product.)
- Obtaining a 100% renewable energy supply for our headquarters, which we hope to achieve via roof-top solar panels.
- Reducing solid wastes from our products and their packaging 25% by 2015.
- Assuring that 100% of our value chain water use is sustainable by 2020.

Source: Company website

Key principle

Setting explicit and challenging organizational sustainability goals provides clear direction to a firm's management and employees.

Strategy – how we will get there

In the field of management, strategy has traditionally been defined as the adoption of courses of action and the allocation of resources necessary for achieving organizational goals (Chandler, 1962). Figure 2.2 illustrates that a firm's strategy answers the question "How do we achieve the organization's mission and goals?" Therefore an organization's sustainability strategy should identify the actions the firm will take to achieve its sustainability mission and goals.

Although many companies feel a sense of urgency just to do something, they often fail to link their sustainability efforts to their business strategy. In a global survey of more than 1500 corporate executives about their perspectives on sustainability and business strategy, Berns *et al.* (2009) found that a majority of respondents believe sustainability is becoming increasingly important to business strategy, and that the risks of failing to act on sustainability are growing. Porter and Kramer (2006: 4) contend that the pressure companies feel to implement sustainability practices too often results in a jumble of uncoordinated sustainability activities, "disconnected from the firm's strategy, that neither make any meaningful social impact nor strengthen the firm's long-term competitiveness." Likewise, Jacopin and Fontrodona (2009) assert that many organizations that have set up a "corporate responsibility" department have done a poor job of aligning that function's activities with firm strategy.

If a firm's sustainability efforts are to provide long-term value to both the company and society, sustainability must be integrated into the firm's strategy. In that regard, Porter and Kramer (2006) advocate that each firm should identify the distinct set of societal issues that it is best equipped to help solve, and from which it can gain the greatest competitive benefit. The pursuit of sustainability may have multiple benefits. Many firms are pursuing a differentiation strategy based on sustainability. By doing this they are seeking to reap a duel benefit of providing value to society as well as distinguishing the firm from competitors (Castello and Lozano, 2009; Siegel, 2009).

The good news for firms looking to build sustainability into their strategy is the recent proliferation of resources to assist management in this undertaking. For example, conducting an Internet search for "sustainability strategy" produces more than a million results. Likewise, a 2008 study found that almost 70% of companies surveyed currently have, or are in the process of developing, a sustainability strategy (Hoffmann, 2008). Without a doubt, an assortment of activities are included in all these strategies. But how can management determine what type of sustainability strategy is appropriate for their business?

A sustainability strategy does not have to be costly or complicated. In fact, as discussed previously, the first stages of sustainability efforts within a firm are often focused on cost reduction. To get off to an easy start, management can simply focus on basic resource-use reduction, such as energy efficiency or recycling. These sorts of efficiency improvements help the environment by conserving natural resources and reducing emissions (Hoffmann, 2008). Likewise, they also can result in significant cost savings, as well as setting the stage for a

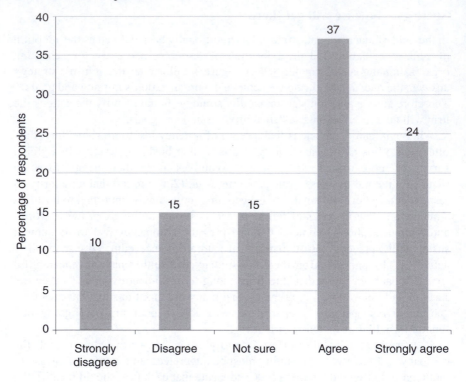

Figure 2.6 Sustainability is included in my company's strategy
(totals more than 100% due to rounding)

firm to develop a more multifaceted and customized approach to sustainability. In our survey (Appendix A), we found that 61% of respondents agreed or strongly agreed with the statement "Sustainability is included in my company's strategy…," whereas 40% of respondents disagreed, strongly disagreed, or were not sure (see Figure 2.6). This result suggests that many firms' leaders have begun to view sustainability as an important part of their company's business model.

Key principle

A well thought out and articulated sustainability strategy will put an organization on the path to achieving its sustainability mission and goals.

Table 2.6 identifies the sustainability strategy at Unilever, a global consumer products company, whose sustainability mission includes four key elements (Unilever, 2011):

Table 2.6 Unilever's sustainability strategy

Strategy	Description
Brand imprint process	Social and environmental considerations are now integrated into the innovation and development plans of major brands.
Engaging consumers	Incorporating consumer research showing that consumers not only want to be reassured that the products they buy are ethically and sustainably produced; they also want to choose brands that are good for them and good for others.
Assessing impacts across the value chain	Commitment to reduced environmental impact across the value chain – from the sourcing of raw materials through production and distribution to consumer use and eventual disposal of residual packaging.
Partnerships	Partners bring expertise on specific issues as well as the networks to deliver practical initiatives on the ground – e.g. UN World Food Programme; World Heart Federation; Global Alliance for Improved Nutrition; FDI World Dental Federation; Global Public–Private Partnership for Handwashing.
UN Global Compact	Signatories to the United Nations Global Compact; committed to living out the Compact's ten principles on human rights, labour, environment and anti-corruption in everyday business operations.

Source: Company website

- We work to create a better future every day.
- We help people feel good, look good and get more out of life with brands and services that are good for them and good for others.
- We will inspire people to take small, everyday actions that can add up to a big difference for the world.
- We will develop new ways of doing business with the aim of doubling the size of our company while reducing our environmental impact.

Sustainability direction setting in action

Direction setting for sustainability begins with the top management of an organization. During the development of this book, we visited and/or spoke with a number of executives who are passionate about their firm's sustainability endeavors. Whether their foundations are a family legacy of stewardship for the land, an entrepreneurial venture developed to help provide a viable future for generations to come, or a large corporation seeking to build lasting value for shareholders and society, each of the people we encountered has focused on building sustainability into their organization's mission, values, goals, and strategy. Here are several examples.

Mission Foods

Headquartered in Irving, TX, the company known as Mission Foods today had its beginnings in 1949 as GRUMA S.A. in Monterrey, Mexico. Since that time, Mission Foods has grown to become one of the largest producers of tortilla products, producing almost a quarter of all tortillas sold in the world, with approximately 5000 employees across fifteen locations throughout the USA.

In 2007, the President of Mission Foods made a conscious decision to pursue sustainability as an overt and focused company-wide effort based on requests from customers such as Walmart and McDonald's. At that time, they appointed a Vice-President of Sustainability, Lucy Gonzalez, to head up the initiative. As part of the direction setting, Mission Foods has established a sustainability slogan – "Today's Mission for a Better Tomorrow" – and a sustainability logo that includes the caption "One Planet, One Mission." Likewise, the stated mission of the company, which includes sustainability, is: "Mission Foods is committed to being a respected category leader in quality tortilla products. We have strong brands, creating profitable growth by exceeding our customers' and consumers' expectations. We measure our success not only in the marketplace, but by the growth of our people and the positive impact we have in the communities we serve." Sustainability has been built into the firm's strategy through various initiatives throughout the company (many of which have been suggested by employees), including internal strategies focused on packaging, energy, solid waste, and water conservation; and external strategies focused on nutrition for children. The next step for Mission Foods is to extend their sustainability efforts outside the company, to suppliers and customers.

Dolphin Blue

Tom Kemper has spent the past eighteen years running Dolphin Blue, Inc. and DolphinBlue.com, a sustainable office products business built on principles of environmental and social sustainability. Dolphin Blue, based in Dallas, TX, was founded by Tom in 1993, based on the belief that we can all be responsible for what we use.

Tom's passion for sustainability stems from his experiences as a boy. In the 1970s, the valley in which Kemper was raised was discovered to have been covered with dioxin-contaminated waste oil. The roads were being sprayed by county contractors to reduce dust in the summer months. A beautiful, spring-fed trout stream – where Kemper fished as a young boy – became suddenly void of life. Rare illnesses and disease began showing up among his former classmates, creating concern among the neighbors in the formerly ecologically vibrant valley. The town was Fenton, Missouri, near the infamous Times Beach, a small town of 2240 residents in St Louis County, Missouri, seventeen miles (twenty-seven kilometers) south-west of St Louis. The town was completely evacuated early in 1983 due to a dioxin scare that made national headlines. It was the largest civilian exposure to dioxin in the United States. This played an important role in the development of Kemper's environmental awareness and activism.

Tom conducted the first public recycling event at the Shakespeare Festival of Dallas in 1992; he collected, sorted, and bagged 350 fifty-gallon bags of recyclable commodities in the three weeks of the festival. He had little success in finding anyone to accept the materials for reprocessing. It was through this exercise that Kemper began to realize the true economics of recycling: "the only way it works, is when we choose products made from materials we attempt to recycle." So he assembled a business and marketing plan for Dolphin Blue in 1993, and the company began providing post-consumer recycled products to its first customers in 1994.

As the founder and CEO of Dolphin Blue, Tom Kemper has set the mission of the company as:

> "a business providing green office supplies and eco-friendly printing for everyday business use. We deliver the highest quality, environmentally responsible office and business products attainable. We educate consumers on the effects of conventional office product consumption. We conduct business always respecting the natural world in which we play a part. We ask our customers to do the same. We treat all with equal regard as fellow occupants of our planet."

Beyond its mission, the firm's values include a commitment to operate and grow in a sustainable manner.

Today, Dolphin Blue is a certified Benefit (B) Corporation. Certified B Corporations use the power of business to solve social and environmental problems. All companies obtaining a B Corp certification must pass a rigorous assessment by the non-profit B Lab group to determine the organizational impact on stakeholders. Certified B Corporations:

- meet rigorous standards of social and environmental performance;
- legally expand their corporate responsibilities to include consideration of stakeholder interests; and
- build collective voice through the power of the unifying B Corporation brand.

As of January 2010, there are over 375 certified B Corporations from over fifty industries, representing a diverse multi-billion dollar marketplace (B Corporation, 2011).

City Garage

Scott Morrison founded City Garage in Plano, TX, in 1993. The company is still privately held, and now has sixteen locations and 105 employees across the Dallas–Fort Worth area, providing auto service and car repairs. Because he is an avid outdoorsman, Scott is passionate about leaving a legacy of environmental stewardship to future generations. He points out that only 8% of waste motor oil in the USA is "re-refined" – a process of turning old motor oil back into a usable product. "In fact," Scott says, "used oil can be re-refined forever, and we should be doing a lot more of it across the country."

As CEO, he has set the sustainability direction of City Garage. Whenever he visits any City Garage location, Scott continually reinforces his commitment to sustainability. He states, "I constantly convey our values to every employee and try to make it personal to them; as many of our people are also outdoorsmen." He continues, "I make sure that we must recycle everything; from office supplies, through motor oil and engine coolants." According to Scott, City Garage returns about 40,000 gallons of used motor oil per year from the cars it services for re-refining. He believes the real issue is to educate consumers to the benefits of re-refined oil. Scott's approach to solving this issue is by educating his employees about the advantages of re-refined oil, so that they can then pass that knowledge on to the consumer.

Montgomery Farm

The community of Montgomery Farm is situated on 500 acres of pristine prairie and forest land in Allen, TX, about twenty miles north of Dallas. Phillip Williams and his sister, Amy Monier, are the visionaries behind the development of Montgomery Farm, which has been in their family for more than sixty-five years. Their goal is "to have Montgomery Farm be a model for the environmentally conscious community of the twenty-first century, by creating a non-polluting, energy-efficient, and sustainable community whose residents live and work in a place of unspoiled beauty." When meeting with Phillip, his passion for the environment and the sustainable development principles of Montgomery Farm is clear. Moreover, the development team he has assembled shares Phillip's commitment to the environment.

Sustainable development is apparent across the community. At the site, greenways throughout the community form a continuum of protected meadows, wildlife preserves, and woodland gardens, all of which are accessible through an extensive network of scenic trails integrated into the pedestrian-friendly neighborhoods of the community. Montgomery Farm also has the first Leadership in Energy and Environmental Design (LEED) Gold-certified home from the US Green Building Council, in Collin County, where Montgomery Farm is located. It is also one of the most energy-efficient homes in the nation, receiving one of the best Home Energy Rating System (HERS) ratings for energy efficiency ever awarded by Energy Star. Moreover, in North Texas, developers typically destroy thousands of trees, some more than 100 years old, to clear land for new neighborhoods and shopping centers. But in Allen, those trees have found a new home nearby. Usually, when you see bulldozers at a home construction site, it's bad news for the trees. At Montgomery Farm, it's a sign of rebirth. For example, the trees being planted at a new home in Montgomery Farm come from different places, including a recently built multi-use facility of shops, restaurants, and apartments, and the construction site for an expansion of a Dallas shopping mall. So far, Montgomery Farm has saved approximately 10,000 trees. Beyond the environmental benefits, the relocation of trees also makes economic sense to Montgomery Farm. Under the city of Allen's tree ordinance, developers pay fees and fines when they destroy large trees. But, by replanting them, Montgomery Farm has saved more than $1.5 million to date.

Key steps to establishing a sustainability direction

1. Why do it? Identify whether your pursuit of sustainability is for risk management, cost reduction, differentiation, or citizenship.
2. Where are we? Determine what you have done so far in sustainability direction setting in your mission, strategy, and values.
3. What can we learn? Identify sources of sustainability information and best practices that the firm can learn from (e.g. publications, competitors, consultants, and even companies from other industries).
4. Who should be involved? Determine who should be included in developing the firm's sustainability direction, such as representatives from key company stakeholders including managers, employees, investors, NGOs, suppliers, and consumers.
5. Where do we need to be? Establish your desired direction regarding sustainability as an organization; set your sustainability mission, values, goals, and strategy.

Summary

*   The process of leading the sustainable organization begins at the macro-/ organization-wide level with the articulation of sustainability as part of the firm's mission, values, goals, and strategy.
*   Strategic management includes strategy formulation (strategic planning), and it also involves strategy implementation and evaluation.
*   A well designed mission statement promotes a set of shared expectations among employees, and communicates a public persona to external stakeholders including the community, consumers, and investors.
*   The exclusion or inclusion of sustainability in a firm's mission is indicative of the firm's commitment to the pursuit of sustainability, and identifies the stage of sustainability development the firm is in – the risk-management, integrated, or citizenship stage.
*   Organizational values refer to beliefs about the types of goals firm members should pursue, as well as ideas regarding standards of behavior organizational members should use to achieve these goals.
*   Goal setting provides the foundation for developing a roadmap of the organization's sustainability activities (the company's sustainability strategy), as well as providing the basis for establishing the metrics that will be used to measure the organization's sustainability progress.

- A firm's strategy answers the question: "How do we achieve the organization's mission and goals?" Therefore an organization's sustainability strategy should identify the actions it will take to achieve its sustainability mission and goals.
- A sustainability strategy does not have to be costly or complicated. To get off to an easy start, management can focus simply on basic resource-use reduction, which will conserve natural resources, reduce emissions, and result in significant cost savings.
- A number of companies have incorporated sustainability into their mission, values, goals, and strategy, including Unilever, Whole Foods, Patagonia, Ecolab, and New Leaf Paper.

Discussion questions

- Does your organization incorporate sustainability into its mission, values, goals, and strategy? If so, how is sustainability articulated in each of these components?
- If your company has not done so yet, how would you incorporate sustainability into your company's mission, values, goals, and strategy?
- What stage of sustainability is your organization in? What about other organizations in your industry?
- If your company is in the first stage of sustainability – risk management – what actions would you suggest for management to move it forward to stage two (integrated)?
- If your company is in the second stage of sustainability – integrated – what actions would you suggest for management to move it forward to stage three (citizenship)?
- If your company is in the third stage of sustainability – citizenship – what actions would you suggest for management to continue to advance the firm's sustainability efforts?
- In general, do you feel that your personal sustainability values "fit" the sustainability values of the company? If so, how do you know? If not, why is there a mismatch?
- In general, do you feel that the sustainability values of the other employees in your firm "fit" the sustainability values of the company? If so, how do you know? If not, why is there a mismatch?

Key tool

Sustainability direction setting assessment

Completing the following scorecard will provide a quick view of the degree to which a firm's mission, values, goals, and strategy are being employed to set the direction for the firm's sustainability efforts.

Table 2.7 Sustainability direction setting assessment

Component	Rating (0 = poor, 10 = excellent)	Notes/ rationale
1 Sustainability is included in the company's mission		
2 The firm's sustainability mission is clear, concise, and repeatable by all employees		
3 Sustainability is included in the company's values		
4 The sustainability values of the firm's workforce fit the sustainability values of the firm		
5 Sustainability is included in the company's goals		
6 The firm's sustainability goals are clear, achievable, challenging, and measurable		
7 Sustainability is included in the company's strategy		
8 The firm's sustainability strategy is clear and understandable by all employees		
9 A cross-section of stakeholders is included in developing and revising the firm's sustainability mission, values, goals, and strategy		
10 The firm's leadership regularly demonstrates the organization's sustainability mission, values, goals, and strategy through their day-to-day behaviors		
Total score		

Steps to complete the assessment

1. Rate each item on a scale of 0 (poor) to 10 (excellent).
2. Make notes for each item to explain the rationale for the numerical rating.
3. Add all ten scores to obtain a total score (maximum = 100).

Rating scale

* 0–20 = poor (significant improvement needed across most or all components)
* 21–40 = below average (improvement needed in several components)
* 41–60 = average (identify areas of weakness and adjust)
* 61–80 = above average (identify areas that can still be improved)
* 81–100 = excellent (continuously review and refine each component as the firm's sustainability efforts evolve)

References

Andersen, T.J. (2000) "Strategic planning, autonomous actions and corporate performance," *Long Range Planning*, 33(2): 184–200.

B Corporation (2011) "About B Corp," www.bcorporation.net/about

Berns, M., Townend, A., Khayat, Z., Balagopal, B., Reeves, M., Hopkins, M.S., and Kruschwitz, N. (2009) "Sustainability and competitive advantage," *MIT Sloan Management Review*, 51(1): 19–26.

Castello, I. and Lozano, J. (2009) "From risk management to citizenship corporate social responsibility: analysis of strategic drivers of change," *Corporate Governance*, 9(4): 373–385.

Chandler, A.D. (1962) *Strategy and Structure*, MIT Press, Cambridge, MA.

Doyle, P. (1994) "Setting business objectives and measuring performance," *Journal of General Management*, 20(2): 1–19.

Etzioni, A. (1960) "Two approaches to organizational analysis: a critique and suggestion," *Administrative Science Quarterly*, 5: 257–278.

Hamel, G. (1996) "Strategy as revolution," *Harvard Business Review*, 74(4): 69–82.

Hargett, T.R. and Williams, M.F. (2009) "Wilh. Wilhelmsen Shipping Company: moving from CSR tradition to CSR leadership," *Corporate Governance*, 9(1): 73–82.

Hoffmann, M. (2008) "Focus on the basics for sustainability strategy," *Plant Engineering*, 62(8): 64.

Jacopin, T. and Fontrodona, J. (2009) "Questioning the corporate responsibility (CR) department alignment with the business model of the company," *Corporate Governance*, 9(4): 528–536.

Kristof-Brown, A., Zimmerman, R., and Johnson, E. (2005) "Consequences of individuals' fit at work: a meta-analysis of person–job, person–organization, person–group, and person–supervisor fit," *Personnel Psychology*, 58(2): 281–342.

Logsdon, J. and Wood, D. (2002) "Business citizenship: from domestic to global level of analysis," *Business Ethics Quarterly*, 12(2): 155–187.

Miller, C.C. and Cardinal, L.B. (1994) "Strategic planning and firm performance: a synthesis of more than two decades of research," *Academy of Management Journal*, 37(6): 1649–1665.

Mintzberg, H. (1987) "Crafting strategy," *Harvard Business Review*, 65(4): 66–75.

Mirvis, P. and Googins, B. (2006) "Stages of corporate citizenship," *California Management Review*, 48(2): 104–126.

Morsing, M. and Oswald, D. (2009) "Sustainable leadership: management control systems and organizational culture in Novo Nordisk A/S," *Corporate Governance*, 9(1): 83–99.

Munilla, L. and Miles, M.P. (2005) "The corporate social responsibility continuum as a component of stakeholder theory," *Business and Society Review*, 110(4): 371–387.

Pekar, Jr, P. and Abraham, S. (1995) "Is strategic management living up to its promise?," *Long Range Planning*, 28(5): 32–34.

Porter, M.E. and Kramer, M.R. (2006) "Strategy and society: the link between competitive advantage and corporate social responsibility," *Harvard Business Review*, 84(12): 78–92.

Posner, B.Z. (2010) "Another look at the impact of personal and organizational values congruency," *Journal of Business Ethics*, 97(4): 535–541.

Ransom, P. and Lober, D.J. (1999) "Why do firms set environmental performance goals? Some evidence from organizational theory," *Business Strategy and the Environment*, 8(1): 1–13.

Rigby, D. (2003) "Management tools survey 2003: usage up as companies strive to make headway in tough times," *Strategy & Leadership*, 31(5): 4–11.

Rok, B. (2009) "Ethical context of the participative leadership model: taking people into account," *Corporate Governance*, 9(4): 461–472.

Rue, L.W. and Ibrahim, N.A. (1998) "The relationship between planning sophistication and performance in small businesses," *Journal of Small Business Management*, 36(4): 24–32.

Schein, E. (1993) *Organizational Culture and Leadership, Classics of Organization Theory*, Harcourt College Publishers, Fort Worth, TX.

Seventh Generation (2011) "Mission and values," www.seventhgeneration.com/press/press-kit

Siegel, D.S. (2009) "Green management matters only if it yields more green: an economic/strategic perspective," *Academy of Management Perspectives*, 23(3): 5–16.

Thompson, J.D. (1967) *Organizations in Action*, McGraw-Hill, New York.

Unilever (2011) "Our sustainability strategy," www.unilever.com/sustainability/introduction/vision/index.aspx

Welch, J. and Welch, S. (2005) *Winning*, HarperCollins, New York.

Wheelen, T.L. and Hunger, J.D. (2008) *Concepts in Strategic Management and Business Policy*, 11th edn, Pearson Education, Upper Saddle River, NJ.

Wilson, I. (1994) "Strategic planning isn't dead – it changed," *Long Range Planning*, 27(4): 20–32.

Zadek, S. (2004) "The path to corporate responsibility," *Harvard Business Review*, 82(12): 125–132.

3   Setting the context
The HR value chain

Once a direction has been set, the process of leading the sustainable organization continues at the macro/organization-wide level with the inclusion of sustainability as part of the firm's human resources (HR) value chain (see Figure 3.1). The HR value chain supports the firm's sustainability aims with macro-level (organization-wide) systems and processes to attract, retain, engage, and develop a workforce with values and talents that are aligned with the firm's sustainability mission, values, goals, and strategy. The macro-level practices set the tone for engaging the workforce in the firm's sustainability endeavors.

As a firm's sustainability efforts are initiated and evolve, so too must management's approach to human capital. Recognizing that the chief source of value is a workforce that is knowledgeable, flexible, and engaged in a firm's strategy requires a fundamental shift in the concept of value management within a corporation (Bartlett and Ghoshal, 2002; Pfeffer, 2005). In order to build a competitive workforce, successful organizations establish HR management practices that support their core values and desired strategy (Dessler, 1999; Chow and Liu, 2009). The organization's values and strategy provide a foundation on which to build HR practices that support a firm's strategic intent and core values. A firm's chosen strategy also provides a foundation for making day-to-day people-management decisions, such as hiring and firing, job design, training, promotions, communicating, and coaching.

The HR value chain provides a conceptual framework for the connection between a firm's strategic direction, its human capital practices, and its workforce. The HR value chain comprises the integrated set of HR management practices – from the sourcing and hiring of talent, through workforce development and performance management, to employee separation – which engages people in a committed pursuit of a set of core values and chosen strategies. Moreover, as a firm evolves along the spectrum of sustainability stages (as discussed in Chapter 2), there will be corresponding shifts in the firm's mission, strategy, and values (Zadek, 2004; Munilla and Miles, 2005; Mirvis and Googins, 2006; Castello and Lozano, 2009). As these shifts occur, the components of the firm's HR value chain must be revisited. A regular alignment check ensures that the organization's HR management practices stay in synch with the evolving sustainability strategy and values of the firm, and that the organization will continue to benefit from a knowledgeable and engaged workforce.

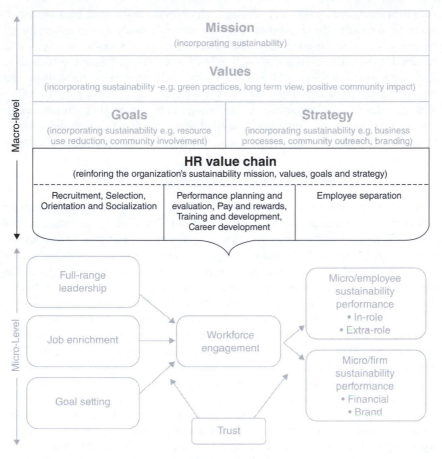

Figure 3.1 Set the context for leading the sustainable organization by including sustainability as part of the firm's HR value chain

---

Key principle

An integrated HR value chain, designed to reinforce the firm's sustainability mission, values, goals, and strategy, sets the stage for engaging a workforce in the firm's sustainability efforts.

Three major stages of the HR value chain

The HR value chain consists of three major stages:

- recruitment, selection, orientation, and socialization;
- continuous reinforcement; and
- employee separation.

As a firm examines its HR value chain, each of these stages must be given equal attention in order to drive a sustainability agenda effectively. For example, hiring people who value sustainability and have the requisite skills to work in a sustainable organization is not enough to build a workforce that can successfully execute a sustainability strategy. Managers and employees throughout the organization also need continuous reinforcement regarding the firm's sustainability strategy, through communications, training, and performance management. Finally, how employee separation is handled communicates to all employees how much the firm values its human capital. Each of these stages is discussed below.

Recruitment, selection, orientation, and socialization

Once the company's sustainability mission, values, goals, and strategy are clearly articulated, the first stage of the HR value chain involves finding and hiring people who fit the desired strategy and stated values. Dessler suggests that companies use value-based hiring practices that screen potential new hires for commitment to a set of chosen values, and reject a large portion of prospective employees. He states, "In many firms the process of linking employees to ideology begins before the worker is even hired" (Dessler, 1999: 23). In fact, there is substantial evidence of a link between a firm's sustainability practices and its attractiveness as an employer (Turban and Greening, 1996; Albinger and Freeman, 2000; Greening and Turban, 2000; Backhaus *et al.*, 2002; Bhattacharya *et al.*, 2008). This attractiveness has been explained by the fact that a firm's sustainability practices enhance its reputation and increase the perceived trustworthiness of an organization for a job-seeker who lacks any previous interaction with the organization (Viswesvaran *et al.*, 1998). This appeal may also be explained using social identity theory. According to this perspective, a firm's sustainability performance sends positive signals to prospective job applicants about what it would be like to work for the company (Greening and Turban, 2000). This attractiveness may help to create a level of commitment to the firm before a candidate is even hired. Beyond potential employees selecting firms that match their personal values, Pfeffer (2005) also advocates selective hiring practices on the part of employers to find employees who "fit" the organization's values.

Key principle

Use value-based hiring practices that screen potential new hires for a commitment to chosen sustainability values.

Continuous reinforcement

As new hires who fit the organization's sustainability strategy and values are brought on board, a process of continuous reinforcement begins. This stage of the HR value chain involves multiple approaches to reinforce the link between the firm's sustainability strategy and its employees. These approaches include incentive pay, information-sharing, empowerment, and skill development (ibid.). Similarly, Dessler (1999) identifies training linked to strategy and values, tradition-building symbols and ceremonies, extensive two-way communications, and promoting the right leaders – those who demonstrate a commitment to the firm's strategy and values – as essential elements of employee engagement. Companies can engage employees in their sustainability strategy and values by consistently emphasizing the sustainability strategy and values of the firm. This can be done by creating community volunteer programs, providing training in sustainability processes, and implementing performance management systems that link the achievement of sustainability goals with compensation (Porter and Kramer, 2006; Ahern, 2009; Lacy *et al.*, 2009; Morsing and Oswald, 2009; Rok, 2009).

Key principle

The HR value chain involves multiple approaches to reinforce the link between the firm's sustainability strategy and its employees.

Employee separation

The final stage of the HR value chain involves the way an organization handles employee separations. Employee separations as a result of layoffs or underperformance are an often-overlooked aspect of employee engagement. Yet how the process of employee separation is handled demonstrates a firm's commitment to being socially responsible through the use of procedures that demonstrate respect for the individuals affected. Every effort should be made to ensure employee separations are ethical and just. When employee separations take place, more individuals than just the exiting employees are attentive to how the process of separation is being handled. The people remaining in the organization also view the way separations are addressed as a clear indicator of the value a firm places on its workforce. In this regard, Dessler (1999) stresses the need for management to demonstrate "organizational justice." When employee separations are conducted in a manner that demonstrates respect for both the individual and the integrity of the organization, people leave with a sense of fairness. This also helps foster a sense of engagement and commitment to the firm among those employees who remain in the organization, because they view the separation process as being fair.

Key principle

How the process of employee separation is handled demonstrates a firm's commitment to being socially responsible through the use of procedures that demonstrate respect for the affected individuals.

The HR function and sustainability

The HR function is frequently ignored as firm leadership identifies and addresses strategic initiatives; recent studies have found little change in the past decade in the tangential status of HR functions (Lawler and Boudreau, 2009; Ulrich *et al.*, 2009). Although HR is traditionally not regarded as a "strategic partner," shifting to a sustainability strategy requires that companies execute their strategy through the firm's human capital (Lacy *et al.*, 2009). Yet many firms have not engaged their workforce in their sustainability efforts through their human capital practices (Porter and Kramer, 2006; Lacy *et al.*, 2009; Sroufe *et al.*, 2010). Moreover, Harmon *et al.* (2010) point out that while developing and executing sustainability strategies should provide HR leaders with an excellent opportunity to improve their strategic role in organizations, they are not taking advantage of that opportunity. Their research found that HR leaders need significant improvement in helping non-HR executives to understand the links between sustainability and talent management, and how HR investments can help drive a sustainability strategy.

In order to help reinforce a sustainability agenda across an organization, HR leaders must work at understanding the strategic aims of the company, how sustainability fits into the firm's strategic direction, and what they can do to reinforce the organization's sustainability efforts through the human capital practices of the firm. The key steps that HR professionals can take to become a strategic partner in the firm's sustainability efforts are as follows.

- Find out how and why management has set the current and past direction (mission, values, goals, and strategy) of the firm.
- Learn how other organizations have utilized their HR value chain to reinforce their strategic initiatives (especially sustainability initiatives), and communicate this learning to firm leadership.
- Work with firm leadership to identify and redesign the human capital practices of the firm that can most easily be changed in the short term (the "quick hits"), in order to begin reinforcing the firm's sustainability efforts (e.g. staffing and hiring, communications, and training).
- Work with firm leadership to identify and redesign the human capital practices of the firm that can be changed over the medium or longer term in order to continue reinforcing the firm's sustainability efforts (e.g. performance management).
- Monitor, assess, and adjust the redesigned human capital practices as they are implemented in order continually to improve the effectiveness of reinforcing the firm's sustainability efforts.

Key principle

The HR function has a key role in designing and implementing an HR value chain that supports the organization's sustainability agenda.

Organizational culture and sustainability

Organizational culture is not easily defined, as demonstrated by the numerous descriptions of the term that have been put forth – ranging from what a firm's management and employees consider to be appropriate business practices (Schein 1985), to the way an organization and its members think about what they do (Bower 2001), to organizational norms, values, beliefs, and attitudes (Goulet and Schweiger 2006). Organizational culture shapes workforce behavior by providing direction, stability, and cohesion to group members (Selznick, 1957). Moreover, individual and group commitment to certain ideals can help create an organizational culture that influences behavior, commitment, organizational decisions, and attitudes (Boeker, 1989).

Components of culture can be isolated, yet no single component of an organization defines its culture (Galpin, 1996). For example, a few managers who continually give orders to their subordinates can be called authoritarian, but the leadership style of these few managers does not create an authoritarian culture. Many other managers within the same organization may frequently solicit input from employees, collectively creating a participatory company culture. Organizational culture is therefore a montage of inter-related elements, with each component of the HR value chain contributing to forming an organization's culture – from the criteria used for hiring and promotion, to the content of communications and how they are delivered, to workforce training, and the performance management process used. In their research, Sroufe *et al.* (2010) found that "leader firms" (those they identified as leading in implementing sustainability strategies) utilized a range of HR systems to reinforce their organization's sustainability efforts in order to build a sustainability culture.

Key principle

Individual components of the HR value chain interact with one another to collectively build an organizational culture that reinforces the firm's sustainability efforts.

Individual HR value chain components and sustainability

Each of the major stages of the HR value chain is comprised of specific human capital practices, including recruiting and hiring, communications, training, leadership behaviors, performance management, decision-making, and so on. Collectively, these processes establish the environment of an organization – its culture – in which employees conduct their day-to-day work. Designed well, each process plays a role in building a workforce that is engaged in the firm's sustainability efforts. The following is a discussion of several individual human capital processes and how they can be used to help establish a culture of sustainability within an organization.

Recruiting and hiring

The first component of a firm's HR value chain with which an employee comes into contact is the company's recruiting and hiring process. The type of people hired, and the values and skills they bring with them into an organization, enable firms to begin building a sustainability culture one hire at a time (ibid.). For example, at a recent roundtable discussion conducted by the recruiting firm Egon Zehnder, which included sustainability executives from leading companies, a chief sustainability officer (CSO) described a recent meeting with his CEO. After listening to a detailed presentation by the CSO about the proposed sustainability direction of the firm, the CEO reacted with, "Too much information. Tell me the one thing we have to do to get this [sustainability] right." Without hesitation, the CSO responded: "Hire the right people" (Lueneburger, 2010: 9). Lueneburger states, "He was not talking merely about hiring a few executives for a sustainability function, but about a substantive culture change that will require the right people with the right competencies across the entire organization."

According to Ehrenfeld, the "right" sustainability competencies include more than just analytical skills. He holds that "Systems thinking is the key to sustainability. Today's [environmental] mess can be blamed on the failure to understand the system within which we are embedded … Unfortunately, systems thinking is given only lip service in most disciplines, especially in MBA programs" (Ehrenfeld, 2010: 10). Likewise, the majority of managers, supervisors, and front-line employees come from single-discipline backgrounds (Galpin *et al.*, 2007), making the systemic perspective required for designing and implementing sustainability efforts more challenging for them to come to terms with. Without consistent, aligned implementation across functional disciplines, however, even the best planned strategy is ineffectual (Aaltonen and Ikavalko, 2002; Allio, 2005).

In our "Current state of sustainability leadership survey" (see Appendix A), we found that only 27% of respondents agreed or strongly agreed with the statement "My company recruits and selects new hires who value sustainability …," whereas 73% of respondents disagreed, strongly disagreed, or were not sure (see Figure 3.2). This finding illustrates that, although many firms' leaders are beginning to view sustainability as being an important part of their companies' strategic direction (mission, values, goals, and strategy), most of these same firms have not yet included sustainability as part of their hiring criteria. Therefore it is evident that much work needs to be done to build sustainability into many organizations' recruiting and hiring practices.

Key principle

The type of people hired, and the values and skills they bring with them into an organization, enable firms to begin building a sustainability culture.

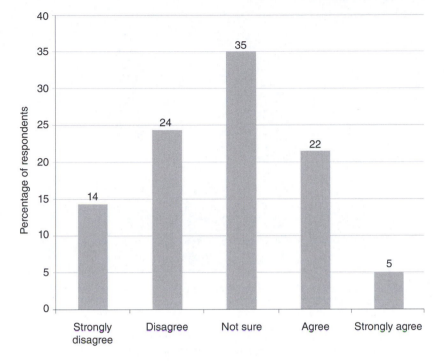

Figure 3.2 My company recruits and selects new hires who value sustainability

Communications

Although the study of organizational strategy (including strategy execution) has been extensive, little attention has been paid to addressing the links between communication and strategy (Forman and Argenti, 2005). Even studies of strategy implementation (e.g. Galbraith and Nathanson, 1978; Lorange, 1982) make communication a secondary concern, focusing instead on issues such as reward systems, organizational structure and processes, and resource allocation. However, the critical role of effective communications in successful strategy execution and employee engagement cannot be overstated. In their discussion of the link between strategy implementation and communications, Saunders *et al.* (2008) point out that a company can build competitive advantage not only through the use of material resources, but by also managing communications so as to improve employee buy-in and help align their interpretations of strategic initiatives.

A number of companies are using internal communication to help create awareness about their sustainability efforts among their workforce. However, more firms need to do a better job of communicating their sustainability strategy to employees. In our survey (Appendix A), we found that almost half (45%) of respondents agreed or strongly agreed with the statement "My company does a good job of communicating to employees and managers the importance of the

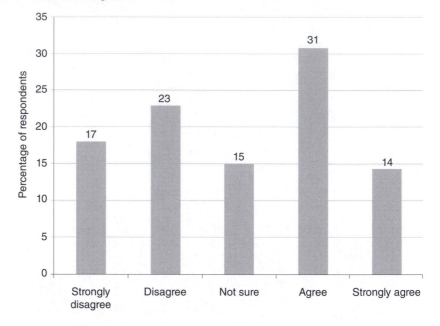

Figure 3.3 My company does a good job of communicating to employees and managers
the importance of the company's sustainability efforts

company's sustainability efforts ...", with just over half (55%) of respondents
indicating that they disagreed, strongly disagreed, or were not sure (see Figure
3.3).

There is no excuse for leaders not communicating their sustainability strategy
to employees. First, technology has transformed communications and access to
information, making it easier to include every employee, even in large
organizations, in a strategic dialogue about the firm's sustainability agenda,
rather than setting goals and ways of achieving them at the top level alone (Rok,
2009). Beyond technology, interpersonal communication is a powerful method
for both sharing information with and soliciting ideas from a firm's workforce.
For example, the health care company Novo Nordisk utilizes interpersonal
communications to spread innovative sustainability approaches and to promote
its sustainability efforts among employees. Novo Nordisk employs approximately
20,000 employees in seventy-eight countries and has its headquarters in
Copenhagen, Denmark. As part of their sustainability strategy, the firm's leaders
have established a team of sixteen high-profile professionals at the holding
company, Novo A/S. Each of the facilitators has a professional background
from senior specialist or managerial positions in Novo Nordisk. They travel in
pairs to visit all business units and levels of the entire organization every third
year, with a main objective of facilitating communication and sharing of
sustainability best practices across the organization (Morsing and Oswald,
2009).

Key principle

Employee communications is a critical component of reinforcing a company's sustainability efforts among the workforce.

Training

Throughout the management literature, training is identified as a key element of engaging a workforce in a firm's strategy execution efforts (e.g. Galpin, 1998; Okumus, 2001; Kundu and Vora, 2004). Organizations that educate their employees about strategic initiatives deliver training most effectively when and where the workforce needs it, often requiring new forms of learning technology, content, and services. For example, a great deal of the training provided by Deloitte Consulting's remodeled learning division consists of blended solutions that combine live classroom training and e-learning (Gold, 2003). Moreover, technology providers such as Microsoft, SAP, IBM, Oracle, Sun, Siebel, and PeopleSoft have added new e-learning functionality (Adkins, 2003). In addition to educating employees, various types of training sessions provide the context in which employees commit themselves to the company and its strategy (Schlesinger and Heskett, 1991).

Although a critical component of strategy execution and workforce engagement, companies appear to be doing a dismal job when it comes to training their workforce about their sustainability strategies. In our survey (Appendix A), we found that only about a quarter (26%) of respondents agreed or strongly agreed with the statement "My company includes sustainability in at least some portion of each employee's and manager's training and development ...", while almost three-quarters (74%) of respondents indicated that they disagreed, strongly disagreed, or were not sure (see Figure 3.4). Likewise, a separate study (Aaron, 2010) found that 75% of employees reported that their firms were not investing in sustainability training, and that many employees wanted sustainability training – only 10% percent of employees felt they had the tools and training needed to help their workplace become more sustainable.

While sustainability training is lacking in many organizations, there are a few bright spots. For example, Aaron (2010) points out that companies including Walmart and Hewlett-Packard have vigorous, well branded employee-engagement programs in order to bring awareness and education about the firms' sustainability efforts to their employees. The concept is to train employees to think like sustainability champions and empower them to find practical, business-building solutions for their workplace. Aaron states, "while I agree ... that some people may be inclined to be more effective in championing sustainability strategies, I vote for using an all-hands-on-deck approach to allow all employees to engage in the process. The more we provide the tools and training and inspire them to action, the quicker we can accelerate the pace to viable solutions" (ibid.: 14).

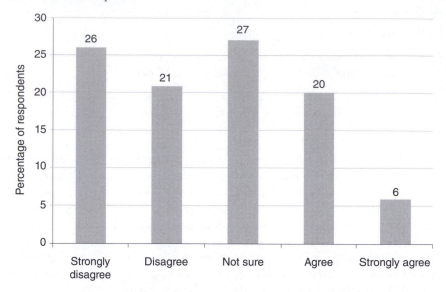

Figure 3.4 My company includes sustainability in at least some portion of each
employee's and manager's training and development

The process of sustainability training should take place early, as new hires are brought into the organization. At Sierra Nevada Brewing Company (SNBC), the new employee-orientation process emphasizes the values and practices of sustainability that characterize the company. SNBC has created two "sustainability coordinator" positions, a full-time manager and a full-time assistant. The sustainability coordinators train new employees on the company's environmental management policies and procedures. SNBC emphasizes sustainability during its new-hire orientation because the company's leaders consider employee–culture fit as being equal to, if not more important than, employee–job fit (Casler *et al.*, 2010). Their rationale is that if an employee is not comfortable with the company's emphasis on sustainability from the outset, then that person will not fit the culture, ultimately compromising his or her productivity and engagement. Moreover, in order continually to enhance the firm's sustainability efforts, as part of its new-hire training SNBC encourages new employees to offer feedback and ideas throughout their tenure with the company (ibid.).

An example of a tailored experiential sustainability learning program comes from the financial services firm HSBC's partnership with the Earthwatch Institute. The program's primary goal is "to inspire action toward sustainability in HSBC offices and communities worldwide" (Barker *et al.*, 2011: 27). A global network of 2200 employees within HSBC, called "climate champions," participate at one of five regional climate centers located near HSBC employment hubs worldwide (USA, UK, Brazil, China, and India). In order to reduce the carbon footprint of the program, climate champions who are accepted into the program participate in a field experience at their closest regional climate center. This local approach also helps the climate champions connect to the sustainability issues that are most relevant to the geographic region in which they live and work. Each day, HSBC's climate champions work next to leading scientists in forest locations, discovering first-hand about the impact of climate change on natural resources. During the evenings, the climate champions participate in workshops, facilitated discussions and training sessions designed to help them become drivers of sustainability in their workplaces and communities. Upon completion of the program, they ask the climate champions to play an active role in creating sustainability-related cost savings, business opportunities, and initiatives within the bank and/or their communities, thus embedding sustainability throughout HSBC's business (Barker *et al.*, 2011).

Key principle

Employee training is a key element of facilitating the effective implementation of a company's sustainability strategy.

Performance planning and evaluation

When it comes to the successful implementation of firms' strategic initiatives, the process by which organizations manage the performance of their employees has been recognized as a crucial component of engaging employees in those initiatives (Galpin, 1997; Caruth and Humphreys, 2008). Michlitsch (2000: 28) notes that "strategy implementation is best accomplished through high-performing people." Therefore it is essential that leaders understand and establish a performance-measurement and feedback system in order "... to link HR management activities with the strategic needs of the business" (Schuler *et al.*, 1991: 389). Managing performance to encourage sustainability efforts requires holding individuals and groups accountable for their contributions by establishing sustainability goals, standards, and norms at individual and organizational levels. This type of accountability is essential to ensure that sustainability goals are not only set, but are also enacted. Accountability at the individual level is ensured by incorporating sustainability targets in annual performance reviews, feedback sessions, regular reporting, professional development, and certifications, as well as rewards and recognition (D'Amato and Roome, 2009).

Although vital to effective workforce engagement in a firm's strategy implementation efforts, companies appear to be doing a poor job when it comes to including sustainability in the performance plans of their managers and employees. In our survey (Appendix A), we found that just over a quarter (27%) of respondents agreed or strongly agreed with the statement "My company requires that sustainability is included within each employee's and manager's performance plan and evaluation ...", while almost three-quarters (73%) of respondents indicated that they disagreed, strongly disagreed, or were not sure (see Figure 3.5). Furthermore, in a case study of ten firms by Pagell and Wu (2009), the

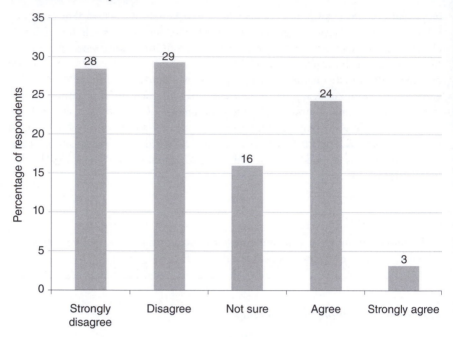

Figure 3.5 My company requires that sustainability is included within each employee's
and manager's performance plan and evaluation

researchers found that nine of the organizations were committed to sustainability,
yet only three had created measurement procedures aimed at encouraging manager
and employee sustainability behaviors. Likewise, Dutta and Lawson (2009) point
out that the performance measurement in many organizations does not consider
sustainability factors that are of strategic importance to the organization. Yet this
is one of the key processes that can effectively motivate a change in corporate
culture to encourage sustainability efforts.

Although sustainability is lacking from employee performance plans within
most organizations, some firms are making progress in this aspect of
reinforcement. For example, General Electric (GE) uses its "Session E" process
formally to review the safety, health, and environmental performance of all
business units. GE has two scorecards: the health-and-safety framework and
scorecard; and the "E-framework", which provides a set of common requirements for GE's environmental programs worldwide. All operations assess their
programs annually against the E-framework, and it is reviewed at Session E.
This tool allows GE to measure performance against a common standard and to
use data to identify areas for improvement (Ramsey and Kruger, 2006). Session
E is an annual occurrence, with GE reviewing the performance of every plant in
a group session that includes the global operational business leader, plant
managers in that business, and the head of corporate environmental programs or

a surrogate. During Session E, plant managers present their results from a standard template. Ramsey and Kruger note, "This is perhaps our most powerful tool in making clear our expectation that operational leaders are responsible for resourcing and managing SH&E as part of their overall responsibilities" (ibid.: 19).

Key principle

Leaders must understand and establish a performance measurement and feedback system that encourages workforce behaviors which support the sustainability efforts of the firm.

Incentives and rewards

It is generally argued that, for performance measurement and evaluation to be motivational, they must be linked with appropriate rewards (Vroom, 1964). Therefore companies must identify, quantify, and reward the impact on the value of the firm that accrues from sustainability efforts. Dutta and Lawson (2009) advocate that the three dimensions of performance – social, environmental, and financial – need to be balanced, and that assigning weights to these three dimensions of performance demonstrates the relative emphasis that management places on each. Companies can and should design their incentive and reward systems to establish, reinforce, and promote the continuous development of a sustainability-oriented organizational culture.

While vital to a firm's successful strategy execution efforts, companies appear to be performing inadequately when it comes to building sustainability into their rewards systems. In our survey (Appendix A), we found that only 19% of respondents agreed or strongly agreed with the statement "My company attaches at least some portion of each employee's and manager's pay and rewards to his/her sustainability performance evaluation …," while 81% of respondents indicated that they disagreed, strongly disagreed, or were not sure (see Figure 3.6). Likewise, in their case study of ten firms, Pagell and Wu (2009) point out that "While nine of the organizations were committed to sustainability, only three had created reward systems that clearly guided behavior toward sustainability goals" (ibid.: 51).

Although currently lacking, various methods can be used to link incentives and rewards to a firm's sustainability agenda. For example, companies can encourage workforce sustainability actions, with incentives such as sharing the time and expenses for volunteer efforts, providing preferential parking for employees driving hybrid cars, or incentivizing employees to use public or company-provided transportation (Dutta and Lawson, 2009). Beyond these incentives, Shell calculates bonuses based on a performance evaluation scorecard, which includes a sustainable development component (Esty and Lauterbach, 2010).

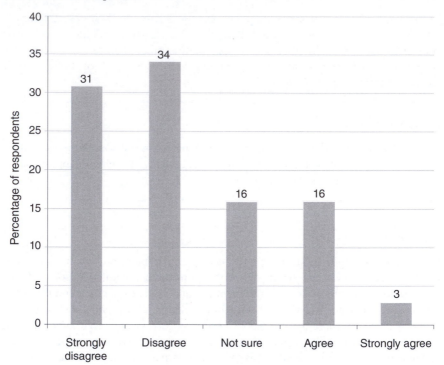

Figure 3.6 My company attaches at least some portion of each employee's and
manager's pay and rewards to his or her sustainability performance evaluation

---

Key principle

For sustainability performance measurement and evaluation to be
motivational, it must be linked with appropriate incentives and rewards.

Employee separation

As mentioned above, the way a firm addresses employee separation (due to layoffs
or individual performance issues) is a frequently overlooked aspect of the HR
value chain that has a recognizable impact on workforce engagement. For instance,
a recent study found that employee engagement was significantly lower if layoffs
had occurred in a company during the preceding twelve months (Kowske *et al.*,
2009). The authors note that organizations may have cut operational costs, but
they are more likely to have at least a portion of their remaining workforce
disengaged, which negatively affects the organization's competitiveness over the
long term. Kowske *et al.* (2009) point out that leaders who ensured that procedures
used to implement separation decisions (such as who stays and who goes) were
perceived as fair (procedural justice) minimized the negative effects of layoffs.

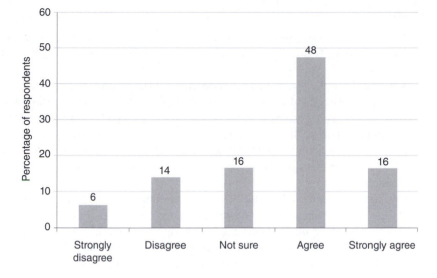

Figure 3.7 My company does a good job of being ethical and fair when it comes to letting people go from the company due to performance issues or layoffs

Likewise, in order to maintain the engagement of survivors (those employees who keep their positions with the firm), Wells (2008) advocates that leaders ensure a perception of fairness in restructuring decisions and to make sure survivors of layoffs know that dismissed employees are treated well.

As highlighted above, our survey (Appendix A) demonstrates that several of the individual processes within the HR value chain seem to be lacking in many firms' sustainability reinforcement efforts. However, we found that the way in which organizations are handling employee separations appears to be a bright spot. We found that almost two-thirds (64%) of respondents agreed or strongly agreed with the statement "My company does a good job of being ethical and fair when it comes to letting people go from the company due to performance issues or layoffs ...", while just over a third (36%) of respondents indicated that they disagreed, strongly disagreed, or were not sure (see Figure 3.7). This finding may be due to the reality that many leaders have had a great deal of experience with layoffs recently due to the economic downturn and/or through their own productivity improvement initiatives over the past several years. Because they have to rely on fewer employees to get the work done, leaders in many firms may have learned the "right" way to handle employee separations in order to ensure those who are kept with the firm stay engaged.

Key principle

Ensuring a perception of fairness in restructuring decisions, and making sure survivors of layoffs know that dismissed employees are treated well, helps to ensure that employees who remain with the firm stay engaged.

Summary

- The HR value chain provides a conceptual framework for the connection between a firm's strategic direction, its human capital practices, and its workforce.
- A firm's HR value chain is the integrated set of HR management practices – from the sourcing and hiring of talent, through workforce development and performance management, to employee separation – which engages people in a committed pursuit of a set of core values and chosen strategy.
- The HR value chain consists of three major stages: recruitment, selection, orientation, and socialization; continuous reinforcement; and employee separation.
- The first stage of the HR value chain involves finding and hiring people that fit the desired strategy and stated values.
- The second stage of the HR value chain involves multiple approaches to reinforce the link between the firm's sustainability strategy and its employees, including incentive pay, information sharing, empowerment, and skill development.
- Although often overlooked, the third and final stage of the HR value chain is the way in which employee separations are handled.
- In order to help reinforce a sustainability agenda across an organization, HR leaders must work at understanding the strategic aims of the company, how sustainability fits into the firm's strategic direction, and what they can do to reinforce the organization's sustainability efforts through the human capital practices of the firm.
- Firms identified as leading in implementing sustainability strategies utilize a range of HR systems to reinforce their organization's sustainability efforts in order to build a sustainability culture.
- Our "Current state of sustainability leadership" survey found that many of the individual processes within the HR value chain seem to be lacking in many firms' sustainability reinforcement efforts; the way in which organizations manage employee separations appears to be the one process that is being conducted well.

Discussion questions

- Does your organization incorporate sustainability into its HR value chain? If so, how is sustainability articulated in each of the individual components of the HR value chain?
- If your company has not done so yet, how would you incorporate sustainability into its HR value chain?
- Does the HR function within your firm do a good job of helping the company's leaders incorporate sustainability into the HR value chain? If so, how? If not, how could HR do a better job of this?

- Does your firm have a sustainability culture? If so, what are some of the key aspects of that culture? If not, what could the firm do to establish a sustainability culture?
- Is sustainability included in your firm's:
  - recruiting and hiring?
  - communications?
  - training?
  - performance planning and evaluation?
  - incentives and rewards?
  - employee separation?

Key tool

Sustainability HR value chain assessment

Completing the following scorecard will provide a quick view of the degree to which a firm's human capital practices are being employed to reinforce the strategic direction of the firm's sustainability efforts.

Steps to complete the assessment

1. Rate each item on a scale of 0 (poor) to 10 (excellent).
2. Make notes for each item to explain the rationale for the numerical rating.
3. Add all ten scores to obtain a total score (maximum = 100).

Rating scale

- 0–20 = poor (significant improvement needed across most or all components)
- 21–40 = below average (improvement needed in several components)
- 41–60 = average (identify areas of weakness and adjust)
- 61–80 = above average (identify areas that can still be improved)
- 81–100 = excellent (continuously review and refine each component as the firm's sustainability efforts evolve)

Table 3.1 Sustainability HR value chain assessment

Component	Rating (0 = poor, 10 = excellent)	Notes/ rationale
1 Sustainability is reinforced through the company's recruiting and hiring practices		
2 Sustainability is regularly reinforced through the company's communications		

Component	Rating *(0 = poor, 10 = excellent)*	Notes/ *rationale*
3 Sustainability is reinforced in the company's training		
4 Sustainability is reinforced through employees' performance management plans (goals and measures)		
5 Sustainability is reinforced through the company's incentives and rewards		
6 Sustainability is reinforced through the company's decision-making processes		
7 Sustainability is reinforced through the company's physical environment (posters/reminders, availability of recycling bins, etc.)		
8 Sustainability is reinforced through the company's management behaviors (our leaders demonstrate sustainability through their actions and words)		
9 Sustainability is reinforced through the company's approach to employee separations (e.g. they are ethical and fair)		
10 The company's human capital practices are regularly updated as the firm's sustainability strategy evolves		
Total score		

References

Aaltonen, P. and Ikavalko, H. (2002) "Implementing strategies successfully," *Integrated Manufacturing Systems*, 13(6): 415–418.

Aaron, S. (2010) "Sustainability: harnessing the collective innovation of all employees," *People & Strategy*, 33(1): 14.

Adkins, S.S. (2003) "The brave new world of learning," *T+D (Training + Development)*, 57(6): 29–37.

Ahern, G. (2009) "Implementing environmental sustainability in ten multinationals," *Corporate Finance Review*, 13(6): 27–32.

Albinger, H.S. and Freeman, S.J. (2000) "Corporate social performance and attractiveness as an employer to different job seeking populations," *Journal of Business Ethics*, 28(3): 243–253.

Allio, M.K. (2005) 'A short, practical guide to implementing strategy," *Journal of Business Strategy*, 26(4): 12–19.

Backhaus, K.B., Stone, B.A., and Heiner, K. (2002) "Exploring the relationship between corporate social performance and employer attractiveness," *Business & Society*, 41(3): 292–318.

Barker, E., Philips, R., Kusek, K., and Thomas, B. (2011) "Earthwatch and HSBC: embedding sustainability into the DNA of HSBC's business," *People & Strategy*, 34(1): 24–32.

Bartlett, C. and Ghoshal, S. (2002) "Building competitive advantage through people," *Sloan Management Review*, 43(2): 34–41.

Boeker, W. (1989) "Strategic change: the effects of founding and history," *Academy of Management Journal*, 32(3): 489–515.

Bower, J.L. (2001) "Not all M&As are alike – and that matters," *Harvard Business Review*, 79(3): 93–101.

Bhattacharya, C.B., Sen, S., and Korschun, D. (2008) 'Using corporate social responsibility to win the war for talent," *MIT Sloan Management Review*, 49(2): 37–44.

Caruth, D.L and Humphreys, J.H. (2008) "Performance appraisal: essential characteristics for strategic control," *Measuring Business Excellence*, 12(3): 24–32.

Casler, A., Gundlach, M.J., Persons, B., and Zivnuska, S. (2010) "Sierra Nevada Brewing Company's thirty-year journey toward sustainability," *People & Strategy*, 33(1): 44–51.

Castello, I. and Lozano, J. (2009) "From risk management to citizenship corporate social responsibility: analysis of strategic drivers of change," *Corporate Governance*, 9(4): 373–385.

Chow, I. and Liu, S.S. (2009) 'The effect of aligning organizational culture and business strategy with HR systems on firm performance in Chinese enterprises," *International Journal of Human Resource Management*, 11: 2292–2310.

D'Amato, A. and Roome, N. (2009) "Toward an integrated model of leadership for corporate responsibility and sustainable development: a process model of corporate responsibility beyond management innovation," *Corporate Governance*, 9(4): 421–434.

Dessler, G. (1999) "How to earn your employees' commitment," *Academy of Management Executive*, 13(2): 58–67.

Dutta, S.K. and Lawson, R.A. (2009) "Aligning performance evaluation and reward systems with corporate sustainability goals," *Cost Management*, 23(6): 15–23.

Ehrenfeld, J.R. (2010) "Sustainability rests in the system, not the product," *People & Strategy*, 33(1): 9–10.

Esty, D.C. and Lauterbach, S. (2010) "Making sustainability part of everyone's job," *People & Strategy*, 33(1): 12–13.

Forman, J. and Argenti, P.A. (2005) "How corporate communication influences strategy implementation, reputation and the corporate brand: an exploratory qualitative study," *Corporate Reputation Review*, 8(3): 245–264,176.

Galbraith, J.R. and Nathanson, D.A. (1978) *Strategy Implementation: The Role of Structure and Process*, West Publishing, St Paul, MN.

Galpin, T. (1996) "Connecting culture to organizational change," *HR Magazine*, 41(3): 84–90.

——(1997). *Making Strategy Work: Building Sustainable Growth Capability*, Jossey-Bass, San Francisco, CA.

——(1998) "When leaders really walk the talk: making strategy work through people," *People & Strategy*, 21(3): 38–45.

Galpin, T.J., Hilpirt, R., and Evans, B. (2007) "The connected enterprise: beyond division of labor," *Journal of Business Strategy*, 28(2): 38–47.

Gold, M. (2003) "Enterprise e-learning," *T+D* (*Training + Development*), 57(4): 28–33.

Goulet, P.K. and Schweiger, D.M. (2006) "Managing culture and human resources in mergers and acquisitions," in G. Stahl and I. Bjorkman (eds) *Handbook of Research in International Human Resource Management*, Edward Elgar, Northampton, MA, pp. 405–432.

Greening, D.W. and Turban, D.B. (2000) "Corporate social performance as a competitive advantage in attracting a quality work force," *Business and Society*, 39(3): 254–280.

Harmon, J., Fairfield, K.D., and Wirtenberg, J. (2010) "Missing an opportunity: HR leadership and sustainability," *People & Strategy*, 33(1): 16–21.

Kowske, B., Lundby, K., Rasch, R., Harris, C., and Lucas, D. (2009) "Turning 'survive' into 'thrive': managing survivor engagement in a downsized organization," *People & Strategy*, 32(4): 48–56.

Kundu, S.C. and Vora, J.A. (2004) "Creating a talented workforce for delivering service quality," *People & Strategy*, 27(2): 40–51.

Lacy, P., Arnott, J., and Lowitt, E. (2009) "The challenge of integrating sustainability into talent and organization strategies: investing in the knowledge, skills and attitudes to achieve high performance," *Corporate Governance*, 9(4): 484–494.

Lawler, E.E. and Boudreau, J.W. (2009) "What makes HR a strategic partner?," *People & Strategy*, 32(1): 14–22.

Lorange, P. (1982) *Implementation of Strategic Planning*, Prentice Hall, Englewood Cliffs, NJ.

Lueneburger, C. (2010) "Caveat venditor: how sustainability is shifting the balance of power," *People & Strategy*, 33(1): 9.

Michlitsch, J.F. (2000) "High-performing, loyal employees: the real way to implement strategy," *Strategy and Leadership*, 28(6): 28–34.

Mirvis, P. and Googins, B. (2006) "Stages of corporate citizenship," *California Management Review*, 48(2): 104–126.

Morsing, M. and Oswald, D. (2009) "Sustainable leadership: management control systems and organizational culture in Novo Nordisk A/S," *Corporate Governance*, 9(1): 83–99.

Munilla, L. and Miles, M.P. (2005) "The corporate social responsibility continuum as a component of stakeholder theory," *Business and Society Review*, 110(4): 371–387.

Okumus, F. (2001) "Towards a strategy implementation framework," *Journal of Contemporary Hospitality Management*, 13(7): 327–338.

Pagell, M. and Wu, Z. (2009) "Building a more complete theory of sustainable supply chain management using case studies of 10 exemplars," *Journal of Supply Chain Management*, 45(2): 37–56.

Pfeffer, J. (2005) "Producing sustainable competitive advantage through the effective management of people," *Academy of Management Executive*, 19(4): 95–106.

Porter, M.E. and Kramer, M.R. (2006) "Strategy and society: the link between competitive advantage and corporate social responsibility," *Harvard Business Review*, 84(12): 78–92.

Ramsey, S. and Kruger, K. (2006) "Safety, health & the environment at GE," *Professional Safety*, 51(12): 18–19, 54.

Rok, B. (2009) "Ethical context of the participative leadership model: taking people into account," *Corporate Governance*, 9(4): 461–472.

Saunders, M., Mann, R., and Smith, R. (2008) "Implementing strategic initiatives: a framework of leading practices," *International Journal of Operations & Production Management*, 28(11): 1095–1123.

Schein, E.H. (1985) *Organizational Culture and Leadership*, Jossey-Bass, San Francisco, CA.

Schlesinger, L.A. and Heskett, J.L. (1991) "The service-driven service company," *Harvard Business Review*, 69(5): 71–81.

Schuler, R.S., Fulkerson, J.R., and Dowling, P.J. (1991) "Strategic performance measurement in multinational corporations," *Human Resource Management*, 30(3): 365–392.

Selznick, P. (1957) *Leadership in Administration: A Sociological Interpretation*, Row, Peterson & Co., Evanston, IL.

Sroufe, R., Liebowitz, J., and Sivasubramaniam, N. (2010) "Are you a leader or a laggard? HR's role in creating a sustainability culture," *People & Strategy*, 33(1): 34–42.

Turban, D.B. and Greening, D.W. (1996) "Corporate social performance and organizational attractiveness to prospective employees," *Academy of Management Journal*, 40: 658–672.

Ulrich, D., Brockbank, W., and Johnson, D. (2009) "The role of strategy architect in the strategic HR organization," *People & Strategy*, 32(1): 24–31.

Viswesvaran, C., Deshpande, S.P., and Milman, C. (1998) "The effect of corporate social responsibility on employee counterproductive behavior," *Cross Cultural Management*, 5(4): 5–12.

Vroom, V. (1964) *Work and Motivation*, Wiley, New York.

Wells, S.J. (2008) "Layoff aftermath," *HRMagazine*, 53(11): 36–41.

Zadek, S. (2004) "The path to corporate responsibility," *Harvard Business Review*, 82(12): 125–132.

Part II
Implementation

4 Engaging employees through full-range leadership

Engagement and the sustainable organization

Successful development and implementation of sustainability initiatives require that the entire organization be fully engaged in the effort. In this chapter, we begin with a definition of engagement, followed by a discussion of the consequences of engagement for organizations. Then we show the critical link between engagement and sustainability. Finally, we introduce full-range leadership as an important antecedent to engagement.

Engagement has recently emerged as a critical element of successful organizations. Employee engagement refers to "the individual's involvement and satisfaction with, as well as enthusiasm for, work" (Harter *et al.*, 2002: 269). Employee engagement has a positive relationship with productivity, profitability, employee retention, safety, and customer satisfaction (Buckingham and Coffman, 1999). The Gallup organization has been on the leading edge of tracking employee engagement, and identifies three levels of engagement. At level 1, engaged employees work with passion and feel a profound connection to their company. They drive innovation and move the organization forward. Level 2 employees are not engaged and have essentially "checked out." They're sleepwalking through their work day, putting time – not energy or passion – into their work. Finally, there are some employees who are actively disengaged. These employees aren't just unhappy at work: they're busy acting out their unhappiness. Every day, these workers undermine what their engaged co-workers accomplish.

Based on extensive, long-term research, the Gallup organization has developed a validated measure of employee engagement consisting of twelve key questions:

1. Do you know what is expected of you at work?
2. Do you have the materials and equipment you need to do your work right?
3. At work, do you have the opportunity to do what you do best every day?
4. In the past seven days, have you received recognition or praise for doing good work?
5. Does your supervisor, or someone at work, seem to care about you as a person?
6. Is there someone at work who encourages your development?

7.   At work, do your opinions seem to count?
8.   Does the mission/purpose of your company make you feel your job is important?
9.   Are your associates (fellow employees) committed to doing quality work?
10. Do you have a best friend at work?
11. In the past six months, has someone at work talked to you about your progress?
12. In the past year, have you had opportunities at work to learn and grow?

Using this measure, Gallup has determined that fewer than 30% of the corporate workforce are truly engaged in their work (Holly and Clifton, 2009). That's fewer than 30% of employees who work with passion and feel a profound connection to their company. Yet employee engagement leads to increased customer engagement, which leads to real revenues and, eventually, more job opportunities for others.

Likewise, the Society for Human Resource Management has identified the level of engagement as a determinant of whether people are productive and stay with the organization – or move to the competition. Without a workplace environment that promotes employee engagement, turnover will increase and efficiency will decline, leading to low customer loyalty and decreased stakeholder value. Ultimately, because the cost of poor employee engagement will be detrimental to organizational success, it is vital for HR to foster positive, effective people-managers along with workplace policies and practices that focus on employee well-being, health, and work/life balance.

Although the link between employee engagement and productivity may be difficult to quantify in financial terms, there are many examples where performance on sustainability issues is leading to higher employee retention rates. Better retention can, in turn, reduce the cost of recruitment and retraining, and can protect a company against the loss of corporate knowledge and experience that is vital for an ongoing commitment to sustainability (Lacy *et al.*, 2010).

A global survey conducted by Accenture (ibid.) highlighted the need for employees at all levels of the organization to become fully engaged in the organization's sustainability efforts. This issue was identified by CEOs as the most important issue facing firms that are seeking to successfully execute their sustainability strategies. An overwhelming 96% of CEOs surveyed reported that for sustainability efforts to succeed, they must permeate all levels of the organization, including people, capability, processes, and systems.

A full-organization approach to sustainability is also encouraged by Haugh and Talwar (2010). They suggest four critical components of an effective effort to embed sustainability across the organization. First, efforts cannot be restricted to the top management team. Second, raising awareness of sustainability issues and initiatives must be cross-functional and spread across the full range of business functions. Third, employees must be provided substantial opportunities to gain practical experience of sustainability initiatives. This allows employees to increase not just their knowledge, but also their interest in and commitment to sustainability; embedding sustainability should include technical and action learning opportunities. Finally, sustainability must be integrated into the long-term learning strategy of the organization. The learning cycle requires opportunities for social learning and expansion of company knowledge systems.

The need for an organization-wide effort requires what Benn *et al.* (2010) call distributed leadership. They suggest that garnering ideas from across the organization is a more important aspect of the CEO's role than considerations of control or authority. CEOs in their survey see the importance of promoting sustainability as a collective quality – something to be shared rather than linked to the influence of any one powerful individual. Organizations that were inclined towards corporate social responsibility encouraged the creation of distributed leadership that allows for dynamic interactions with a wide range of stakeholders, rather than leadership just at the corporate level.

According to the Accenture report, managers at every level need to acknowledge that awareness of sustainability helps them to understand better the needs and concerns of society. This helps minimize risks and can actually generate business opportunities (Lacy *et al.*, 2010). In order for this transition from employees merely acknowledging the importance of sustainability to the point at which such issues are incorporated in their day-to-day work remains a key challenge. This shift can be supported by the incorporation of sustainability objectives into employees' performance assessments.

As described in Chapters 1–3 of this book, creating a sustainable organization requires a comprehensive effort that addresses both the macro-level aspects of a firm's mission, values, goals, and strategy, and the HR value chain, as well as the micro-level issues associated with employee leadership, task design, and performance management. The macro-level organizational practices that make up the HR value chain provide the context in which an engaged workforce can develop, and they have been associated with organizational success in sustainability efforts (Ahern, 2009; Lacy *et al.*, 2009; Morsing and Oswald, 2009; Rok, 2009). But these macro-level practices alone will not create the high levels of employee engagement necessary to achieve an organization's sustainability objectives. In fact, the best intentioned sustainability strategies may be undermined by the actions of first-line managers. Thus the manner in which an employee perceives and responds to their task and their manager will ultimately determine the impact of the organization's macro-level practices.

The macro-level organizational practices that make up the HR value chain provide the context in which an engaged workforce can develop. However, if these organizational practices are not reinforced at the micro-level, employee engagement may still be lacking. Bedeian and Armenakis (1998: 59) state that, "As assuredly as Gresham's Law states that bad money drives out good money, incompetent managers, wherever situated, inevitably drive away good employees." This is reinforced by Buckingham and Coffman (1999), who observed that "employees don't quit companies, they quit managers."

In the Chapters 4–6, we focus on the micro-level leadership components of the Leading the Sustainable Organization model (Figure 4.1) that contribute to high levels of employee engagement in a firm's sustainability efforts. In this chapter,

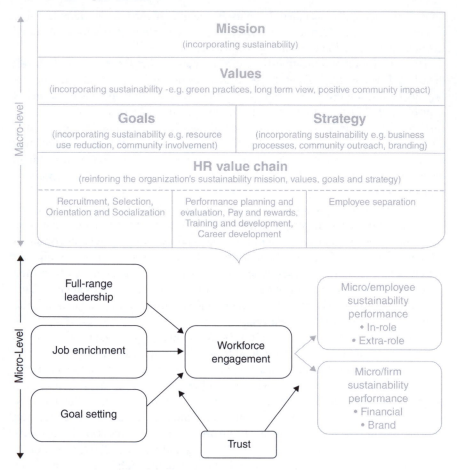

Figure 4.1 Micro-level leadership components

we focus on the full-range leadership model. Chapter 5 discusses the role of task design and performance management in engaging the workforce in the organization's sustainability effort. In Chapter 6, we focus on the quality of the leader–follower relationship and the importance of trust in engaging employees.

All too frequently, discussions of leadership in the context of sustainability focus solely on the CEO and the top management team. However, as the above discussion indicates, successful implementation of a sustainability strategy will require leadership at all levels of an organization. Because these leaders must articulate the vision and strategy, as well as show employees how their individual role contributes to the organization's sustainability effort, we recommend that leaders utilize a full range of leadership behaviors. The full-range leadership model is discussed in the remainder of this chapter.

> **Key principle**
>
> Successful sustainability efforts require the engagement of the entire organization, at all levels.

Full-range leadership

Over the past three decades, the social scientific approach to leadership has been dominated by the "full-range of leadership" paradigm that encompasses both transactional and transformational leadership behaviors (Avolio, 1999; Bass and Riggio, 2006). The full-range model has its roots in the pioneering work of James MacGregor Burns (1978). In his seminal book, *Leadership*, he defined leadership as "leaders inducing followers to act for certain goals that represent the values and the motivations – the wants and needs, the aspirations and expectations – *of both leaders and followers*" (ibid.: 19, original italics). Burns suggested that the interaction between a leader and his/her followers could take two fundamentally different forms: transactional and transforming. These two forms of leadership were further developed by Bass and his associates (Bass, 1985; Bass and Avolio, 1994; Avolio, 1999) who operationalized them in the multifactor leadership questionnaire.

According to the full-range view, the highest level of effectiveness is achieved when leaders engage in a two-stage process in which transactional leadership provides the basis for the subsequent development of transformational leadership (Avolio, 1999). Transactional leadership behavior focuses on the leader's efforts to clarify performance expectations and the rewards that may be expected for meeting these expectations. Leaders who fail to establish this set of role expectations leave followers with an ill-defined sense of direction and ambiguous task assignments. However, when done correctly, the clarification of role expectations provides the basis for more mature relationships between a leader and his/her followers to evolve over time. When the leader consistently follows through with the rewards that are promised in exchange for that performance, trust and commitment emerge.

Transactional leadership occurs when a leader exchanges something of economic, political, or psychological value with a follower. These exchanges are based on the leader identifying performance requirements and clarifying the conditions under which rewards are available for meeting these requirements. The goal is to enter into a mutually beneficial exchange, but not necessarily to develop an enduring relationship. Although a leadership act transpires, it is not one that binds the leader and follower together in a mutual and continuing pursuit of a higher purpose.

Transactional leadership is contrasted with transforming leadership, which occurs when individuals engage with each other in such a way that the leader and follower raise one another to higher levels of motivation and morality. Effective

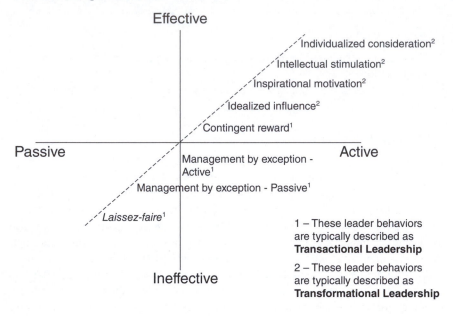

Figure 4.2 The full range of leadership

transformational leaders may exhibit transactional behaviors, but their leadership style also includes one or more of the following characteristics: idealized vision, inspirational motivation, intellectual stimulation, and individualized consideration (Bass, 1985; Bass and Avolio, 1994). These characteristics are assumed to transform followers and motivate them to do more than initially expected. This transformation presumably occurs through raising followers' awareness of the significance of designated outcomes, getting followers to transcend their self-interests for the good of the organization, or augmenting followers' needs on Maslow's (1954) hierarchy of needs (Bass, 1985). Although leaders' and followers' purposes may begin as separate but related, they eventually become fused into a linkage of power bases that provides support for both members of the relationship.

The research that has focused primarily on transformational dimensions of the full-range leadership model has produced impressive results (Podsakoff *et al.*, 1990; Yammarino and Bass, 1990; Howell and Avolio, 1993; Bycio *et al.*, 1995; Whittington *et al.*, 2004). There is significant support for a positive relationship between transformational leadership behavior and a number of follower outcomes.

The full-range leadership model is a primary tool for developing the relationship between the company and its staff, and is critical in the process of adopting sustainability practices throughout the organization. Therefore it is important to understand each set of leader behaviors that comprise full-range leadership – transactional and transformational – in order to grasp the benefits of the full-range model. There are clear differences in the types of behavior associated with transactional and transformational leadership (see Figure 4.2).

Transactional leadership

Transactional leadership emphasizes two factors: contingent reward and management-by-exception (Bass, 1985). Contingent reward refers to the efforts made by the leader to clarify expectations so that followers will understand what they need to do in order to receive rewards. Therefore contingent reward shows the degree to which you let others know what they need to do in order to be rewarded, emphasizing what you expect from them, and that you recognize their accomplishments. The key characteristics of contingent rewards include:

- clarifying expectations about the job and the results you expect;
- answering the question: "what does a good job look like?";
- creating a clear line of sight between results and rewards;
- not promising rewards you cannot deliver; and
- keeping your promises.

Management-by-exception is a less active approach to leadership than contingent rewards, which essentially informs followers of job expectations, but resists further involvement with the follower unless the follower's actual performance varies significantly from those expectations. Transformational leadership includes behaviors associated with inspirational motivation (e.g. articulating a vision that followers find meaningful and challenging), ascribed charisma (e.g. role modeling behaviors that gain admiration and trust), intellectual stimulation (e.g. encouraging followers to question assumptions and think "out of the box"), and individualized consideration (e.g. empowering, supporting, and paying attention to the needs of followers). These behaviors describe leaders with strong social skills who are capable of communicating effectively in order "to arouse, inspire, and motivate followers (Riggio *et al.*, 2003: 85).

Researchers contend that differences in leadership behaviors are based on the cognitive processes that underlie those behaviors (Wofford and Goodwin, 1994; Wofford, *et al.*, 1998; Goodwin, *et al.*, 1999). According to Wofford and Goodwin's cognitive interpretation of leadership, "the cognitive processes and structures that are primed by feedback and the leader's environment influence the use of transactional and transformational leadership behaviors when they are within the leader's repertoire of possible behaviors" (Wofford and Goodwin, 1994: 163).

Using an information-processing framework (Lord and Maher, 1991), Wofford and Goodwin (1994) developed several propositions about the use of transactional leader behaviors. They discussed the content of leader schemata and suggested that transactional leaders focused on clearly defined performance goals. The content of transactional leader schemata and scripts is expected to emphasize goal difficulty, goal commitment, task-related knowledge and skills, role expectations, and incentives that relate to individual employee or organizational sub-unit performance. On the other hand, transformational leaders should be more likely to possess cognitive structures containing information about organizational vision accomplishment, development of long-term relationships, and more abstract, "big-picture" concepts. Consistent with the general propositions of Wofford and Goodwin's theory, Wofford *et al.* (1998) found support for different cognitive processes underlying transactional and transformational leadership in a field study. Transactional leaders possessed schemata in their cognitive content that represented transactional behavior, and transformational leaders possessed transformational content in their memory structures.

Despite transformational leaders' primary focus on an overall "vision," Wofford and Goodwin (1994) argued that they also have within their cognitive repertoire elements pertaining to transactional leadership behavior. That is, individuals who are able to engage in transformational leadership behavior may revert to the more concrete level of *quid pro quo* agreements and engage in transactional leadership behavior as needed. Thus they argued for a hierarchical framework, progressing upward from concrete to more abstract representations in memory, wherein transactional leadership provides the more concrete and pragmatic foundation upon which transformational leadership rests.

The idea that transactional and transformational leadership behaviors form a hierarchy is also consistent with Avolio's (1999) work on the full-range model of leadership. The essence of this framework is that effective leaders engage in a full range of behaviors that encompass elements of both transactional and transformational leadership, with transactional leadership as the basis for the subsequent development of transformational leadership. Leaders who lack a foundation of transactional leadership are often likely to leave their employees' role expectations unclear, which results in an ill-defined sense of direction and ambiguous task assignments. When these role expectations have been appropriately clarified through the use of transactional leadership behavior, however, more mature relationships between a leader and his/her followers can evolve over time. Thus the clarification of role expectations provides a crucial basis for building a more general framework of mutual expectations between leader and follower. Furthermore, when leaders honor their various transactional arrangements with their followers, trust begins to develop, creating the foundation for a sustained relationship that enables the effective utilization of the full range of leadership behaviors (ibid.).

When used correctly, transactional behaviors can accomplish leader's goals and also satisfy the interests of followers. These behaviors can take either of two forms. **Constructive transactions** are those that are used to clarify expectations and identify the linkages between performance and rewards. If done properly, these exchanges form a "compact of expectations" (ibid.: 36) by which followers will evaluate the consistency and trustworthiness of their leader. In contrast, **corrective transactions** focus on creating a desired change in behavior, cooperation, or attitude. These transactions are somewhat negative in that they clarify what must be done to avoid censorship, reproof, punishment, or other disciplinary actions (ibid.).

Both constructive and corrective transactions are important to the effectiveness of transactional leaders. As they honor constructive agreements and consistently apply corrective measures, their followers are able to develop perceptions about the consistency of their behavior and the likelihood that they will meet their leaders' expectations. As such, the recognition of transactional behaviors by followers is important to the basis for a productive, trusting relationship. Transactional leadership is generally easily identifiable because the behaviors revolve around key issues of employment such as wages/salaries, performance feedback, rewards for performance such as promotions, etc., and are centered on relatively concrete acts. Although transactional leadership is not enough to develop the full potential of followers, it is a necessary transitional step in developing the trust between a leader and follower that is required for transformational leadership to be implemented and become effective (ibid.). Yet, in the process, followers are able to receive the benefit of the transactional leader's guidance. Thus followers will benefit not only from their association with a transformational leader, but also from their relationship with a transactional leader.

In a widely supported theory of performance, Locke and Latham (1990) show that when employees have challenging and specific goals, suitable task strategies, and clear linkages between performance and the rewards that they desire, high levels of performance will result. Because effective transactional leaders clarify the performance expectations they hold for their followers, these followers can be expected to perform well. Furthermore, when followers are confident about their specific role expectations, they may be more likely to go beyond the formal aspects of in-role performance and engage in extra-role behaviors, such as organizational citizenship behavior (OCB; Organ, 1988). This refers to behavior of an employee that is discretionary, not rewarded or recognized in an explicit way by the organization, and that tends to promote efficient and effective functioning of the organization (ibid.). OCBs are performed spontaneously by employees who elect to go beyond in-role expectations – voluntary activities performed by employees without regard to possible sanctions or incentives (Organ, 1988, 1990). Consequently, transactional leadership behavior may enable both in-role and extra-role performance.

Another important outcome for organizations is employee commitment, which has been dichotomized into attitudinal and behavioral components (Meyer and Allen, 1991). Mowday *et al.* (1982: 26) have distinguished these components as follows: "Attitudinal commitment focuses on the process by which people come to think about their relationship with the organization … Behavioral commitment, on the other hand, relates to the process by which individuals become locked into a certain organization and how they deal with this problem."

Transactional leadership behavior engages followers in an agreement that specifies followers' performance expectations and the consequences for meeting those expectations. When transactional leaders consistently follow through on these exchange agreements, employees may develop affective commitment to the organization; they continue with the organization because they want to, and they feel comfortable staying because they understand clearly what they must do to receive desired rewards.

Key principle

Transactional leadership is an important dimension of the full-range model of leadership because it clarifies expectations and creates a clear linkage between performance and rewards.

Transformational leadership

Although transformational leaders may exhibit transactional behaviors (Wofford *et al.*, 1998), their leadership style also includes one or more of the following behaviors: idealized influence, inspirational motivation, intellectual stimulation, and/or individualized consideration (Bass and Avolio, 1994; Avolio, 1999). Through the use of these behaviors, followers are transformed by raising their awareness of the significance of designated outcomes. This transformation enables followers to transcend their self-interest for the good of the organization. Leaders who exhibit these transformational behaviors can expect "performance beyond expectations" (Bass, 1985), as well as a wide variety of other positive outcomes in organizational settings.

Idealized influence

Idealized influence refers to the role-modeling behavior of transformational leaders. These leaders consider the needs of others over their own, share risks with their followers, and demonstrate high standards of moral conduct. Leaders demonstrate faith in others by empowering followers and creating a joint sense of mission (Avolio, 1999). Consequently, their followers identify with and attempt to emulate them (Bass and Avolio, 1994; Bass and Riggio, 2006). Therefore idealized influence indicates whether you hold subordinates' trust, maintain their faith and respect, show dedication to them, appeal to their hopes and dreams, and act as their role model. Key characteristics of idealized influence include:

- setting examples for showing determination;
- displaying extraordinary talents;
- taking risks;
- creating a sense of empowerment in followers;
- showing dedication to the cause;
- creating a sense of joint mission;
- dealing with crises;
- using radical solutions;
- engendering faith in others.

Inspirational motivation

Transformational leaders use inspirational motivation to build emotional commitment to the organization's vision or goal. They do this by articulating a vision that portrays an attractive future that provides meaning and challenge for followers (Bass, 1985). Clear expectations are communicated, with a demonstrated commitment to goals and the shared vision. Inspirational motivation entails the degree to which you provide a vision, use appropriate symbols and images to help others focus on their work, and try to make others feel their work is significant. Key characteristics of inspirational motivation include:

- providing meaning and challenge;
- painting an optimistic future;
- molding expectations that create self-fulfilling prophecies;
- thinking ahead;
- taking a first step – often with risk to oneself.

Intellectual stimulation

Transformational leaders are change agents who use intellectual stimulation to question assumptions, reframe problems, and approach existing situations from a fresh perspective (ibid.). This behavior encourages innovation and creativity. Participation and creative risk-taking are encouraged without the fear of public criticism or penalty for departure from the leader's ideas (Heifetz, 1994; Bass and Riggio, 2006). Intellectual stimulation shows the degree to which you encourage others to be creative in looking at old problems in new ways, create an environment that is tolerant of seemingly extreme positions, and nurture people to question their own values and beliefs and those of the organization. Intellectual stimulation incorporates:

- questioning assumptions;
- encouraging followers to employ intuition;
- entertaining ideas that may have seemed silly at first;
- creating imaginative visions;
- asking followers to rework problems they thought they had solved;
- seeing unusual patterns;
- using humor to stimulate new thinking.

Individualized consideration

Individualized consideration refers to the transformational leader's mentoring role, in which the leader pays special attention to each individual's need for personal growth and achievement (Bass, 1985). Transformational leaders use delegation as a developmental tool to advance followers to successively higher levels of potential. They also are intentional about creating learning opportunities and a supportive environment to facilitate followers' development. Individualized consideration indicates the degree to which you show interest in others' well-being, assign projects individually, and pay attention to those who seem less involved in the group. Individualized consideration involves:

- answering followers with minimum delay;
- showing concern for followers' well-being;
- assigning tasks on the basis of individual needs and abilities;
- encouraging two-way exchange of ideas;
- being available when needed;
- constantly encouraging self-development;
- effectively mentoring, counseling, and coaching peers and followers.

Over the past two decades, researchers have consistently found impressive results for the transformational behavior of leaders (Bass, 1985; Bass and Avolio, 1994; Avolio, 1999; Whittington *et al.*, 2004; Bass and Riggio, 2006). Subordinates' performance may be "beyond expectations" (Bass, 1985) when leaders use transformational behaviors; however, other positive outcomes in organizational settings are also related to this leadership style: high satisfaction with the leader (Podsakoff *et al.*, 1990; Bycio *et al.*, 1995); employees' affective commitment to the organization (Bycio, *et al.*, 1995; Whittington *et al.*, 2004); trust in the leader (Podsakoff *et al.*, 1990); and OCBs (Whittington *et al.*, 2004).

Key principle

When transactional leadership is supplemented with transformational leadership behaviors, employee commitment and performance go beyond expectations.

Full-range leadership and sustainability in practice

In the context of sustainability, leaders must utilize a full range of behaviors to engage individual employees in the organization's strategy. In order to do this, leaders must first set clear performance expectations for employees, which incorporate aspects of sustainability into individual performance plans (contingent reward). Leaders must demonstrate their commitment to the strategy and act as role models for incorporating sustainability into the daily activities and operations of the firm (idealized influence). Leaders also need to communicate an inspiring vision that provides a compelling case for sustainability (inspirational motivation). These leaders will also challenge the *status quo* and encourage employees to be creative in finding innovative ways to implement sustainable practices throughout the organization (intellectual stimulation).

While each of these full-range behaviors is important for engaging employees in the organization's sustainability strategy, the results of our survey suggest that these practices have not been fully implemented in the organizations we sampled. Our survey of a wide variety of industries (see Appendix A) found that only 35% felt that their manager had done a good job of communicating a clear vision for incorporating sustainability at the departmental level. The failure to communicate an effective vision is no doubt linked to a low level of commitment to the sustainability effort. Only 35% of respondents reported a high level of perceived commitment to sustainability in their managers. Furthermore, only 33% of our respondents reported that they had specific sustainability-related performance objectives.

Summary

*   Successful development and implementation of sustainability initiatives requires the engagement of the entire organization.
*   The benefits of engagement include increased employee commitment, performance, and satisfaction.
*   The costs associated with disengagement include high turnover of key talent and low levels of commitment to key organizational initiatives.
*   CEOs now recognize that engagement is a strategic imperative.
*   A key antecedent of engagement is the presence of full-range leaders throughout the organization.
*   Transactional leadership cannot be ignored. It sets the foundation of clear expectations.
*   Transformational leadership behaviors augment transactional leadership behavior by inspiring commitment to the organization's sustainability vision and developing employees' capacity to achieve that vision.

Discussion questions

*   How does your organization define engagement?
*   How is engagement measured in your organization?
*   Using the guidelines provided in Table 4.1, how does your organization measure up on the elements of engagement?
*   Is your firm's sustainability effort a top-down only approach, or does it seek to engage the entire organization?
*   Which leadership behaviors seem most prominent in your organization, transactional or transformational?
*   How committed is your organization to developing full-range leaders at every level?

Key tool

Full-range leadership assessment

This questionnaire provides a description of your manager's behavior. Please describe your manager by answering the questions below.

Table 4.1 Full-range leadership assessment

Leadership behaviors	Not at all	Once in a while	Sometimes	Fairly often	Frequently
1 It required a failure to meet an objective for him/her to take action	1	2	3	4	5
2 Focused attention on irregularities, mistakes, exceptions, and deviations from standards	1	2	3	4	5
3 Gave me what I wanted in exchange for my support	1	2	3	4	5
4 Work had to fall below minimum standards for him/her to try to make improvements					
5 Made clear what I could expect to receive if my performance met designated standards	1	2	3	4	5
6 Failed to intervene until problems became serious	1	2	3	4	5
7 Instilled pride in being associated with him/her	1	2	3	4	5
8 Worked out agreements with me on what I would receive if I did what needed to be done	1	2	3	4	5
9 Talked optimistically about the future	1	2	3	4	5
10 Closely monitored my performance for errors	1	2	3	4	5
11 Provided useful advice for my development	1	2	3	4	5
12 Kept track of my mistakes	1	2	3	4	5
13 Went beyond his/her own self-interest for the good of our group	1	2	3	4	5
14 Negotiated with me about what I could expect to receive for what I accomplish	1	2	3	4	5
15 Expressed his/her confidence that we would achieve our goals	1	2	3	4	5
16 Things had to go wrong for him/her to take action	1	2	3	4	5

Leadership behaviors	Not at all	Once in a while	Sometimes	Fairly often	Frequently
17 Focused me on developing my strengths	1	2	3	4	5
18 Provided his/her assistance in exchange for my effort	1	2	3	4	5
19 Provided reassurance that we would overcome obstacles	1	2	3	4	5
20 Provided continuous encouragement	1	2	3	4	5
21 Directed his/her attention toward failure to meet standards	1	2	3	4	5
22 Sought differing perspectives when solving problems	1	2	3	4	5
23 Told me what to do to be rewarded for my efforts	1	2	3	4	5
24 Displayed extraordinary talent and competence in whatever he/she undertook	1	2	3	4	5
25 Problems had to become chronic before he/she would take action	1	2	3	4	5
26 Searched for mistakes before commenting on my performance	1	2	3	4	5
27 Made sure that we receive appropriate rewards for achieving performance targets	1	2	3	4	5
28 Suggested new ways of looking at how we do our jobs	1	2	3	4	5
29 Treated each of us as individuals with different needs, abilities, and aspirations	1	2	3	4	5
30 His/her actions built my respect for him/her	1	2	3	4	5
31 I earned credit from him/her by doing my job well	1	2	3	4	5
32 Talked enthusiastically about what needed to be accomplished	1	2	3	4	5
33 Articulated a compelling vision of the future	1	2	3	4	5

Leadership behaviors	Not at all	Once in a while	Sometimes	Fairly often	Frequently
34 Got me to look at problems from many different angles	1	2	3	4	5
35 Promoted self-development	1	2	3	4	5
36 Encouraged non-traditional thinking to deal with traditional problems	1	2	3	4	5
37 Gave personal attention to members who seemed neglected	1	2	3	4	5
38 Expressed his/her satisfaction when I did a good job	1	2	3	4	5
39 Encouraged addressing problems by using reasoning and evidence, rather than unsupported opinion	1	2	3	4	5

Where an item is irrelevant or does not apply, or where you are uncertain or don't know, leave the answer blank.

Listed below are descriptive statements about your manager. For each statement, we would like you to judge **how frequently** that person displayed the behavior described. Please circle the appropriate number.

Scoring instructions

The full range of leadership behaviors includes both transformational and transactional behaviors.

Transformational scoring

The transformational leadership behaviors are described using five sub-scales: ascribed charisma (AC), individualized consideration (IC), intellectual stimulation (IS), inspirational motivation (INSP), and implicit contingent rewards (IMPCR).

An average score on each sub-scale can be calculated for each dimension of transformational leadership. Any sub-scale score higher than 4 indicates a high level of this particular behavior.

- SUM the scores for AC questions = Q8 + Q13 + Q19 + Q24 + Q30 DIVIDE SUM by 5
- SUM the scores for IC questions = Q11 + Q17 + Q29 + Q35 + Q37 DIVIDE SUM by 5
- SUM the scores for IS questions = Q22 + Q28 + Q34 + Q36 + Q39 DIVIDE SUM by 5

- SUM the scores for INSP questions = Q10 + Q15 + Q20 + Q32 + Q33 DIVIDE SUM by 5
- SUM the scores for IMPCR questions = Q3 + Q18 + Q27 + Q31 + Q38 DIVIDE SUM by 5

A total score for transformational leadership can be determined by adding the sub-scale scores. Total transformational scores of 20 or higher indicate a high level of transformational leadership.

Transactional scoring

The transactional leadership behaviors are described using three sub-scales: explicit contingent reward (EXPCR), management by exception – passive (MBEP), and management by exception – active (MBEA).

An average score on each sub-scale can be calculated for each dimension of transactional leadership. Any sub-scale score higher than 4 indicates a high level of this particular behavior.

- SUM the EXPCR items = 6 + 9 + 14 + 23 DIVIDE SUM by 4
- SUM the MBEP items = 1 + 4 + 7 + 16 + 25 DIVIDE SUM by 5
- SUM the MBEA items = 2 + 5 + 12 + 21 + 26 DIVIDE SUM by 5

A total score for transactional leadership can be determined by adding the sub-scale scores. Total transformational scores of 12 or higher indicate a high level of transactional leadership.

References

Ahern, G. (2009) "Implementing environmental sustainability in ten multinationals," *Corporate Finance Review*, 13(6): 27–32.

Allen, N. and Meyer, J. (1990) "The measurement and antecedents of affective, continuance and normative commitment to the organization," *Journal of Occupational Psychology*, 63: 1–18.

Avolio, B. (1999) *Full Leadership Development*, Sage, Thousand Oaks, CA.

Bass, B. (1985) *Leadership and Performance beyond Expectations*, The Free Press, New York.

Bass, B. and Avolio, B. (1994) *Improving Organizational Effectiveness through Transformational Leadership*, Sage, Thousand Oaks, CA.

Bass, B. and Riggio, R. (2006) *Transformational Leadership*, 2nd edn, Lawrence Erlbaum and Associates, Mahwah, NJ.

Bedeian, A. and Armenakis, A. (1998) "The cesspool syndrome: how dreck floats to the top of declining organizations," *Academy of Management Executive*, 12(1): 58–67.

Benn, S., Todd, L.R., and Pendleton, J. (2010) "Public relations leadership in corporate social responsibility," *Journal of Business Ethics*, 96: 403–423.

Buckingham, M. and Coffman, C. (1999) *First, Break All The Rules: What the World's Greatest Managers do Differently*, Simon & Schuster, New York.

Burns, J. M. (1978) *Leadership*, Harper & Row, New York.

Bycio, P., Hackett, R., and Allen, J. (1995) "Further assessments of Bass's (1985) conceptualization of transactional and transformational leadership," *Journal of Applied Psychology*, 80: 468–478.

Dienesch, R.M. and Liden, R.C. (1986) "Leader–member exchange model of leadership: a critique and further development," *Academy of Management Review*, 11: 618–634.

Goodwin, V.L., Wofford, J.C., and Boyd, N. (1999) "A laboratory experiment testing the antecedents of leader cognitions," *Journal of Organizational Behavior*, 21: 769–788.

Harter, J.K., Schmidt, F.L., and Hayes T.L. (2002) "Business-unit-level relationship between employee satisfaction, employee engagement, and business outcomes: a meta-analysis," *Journal of Applied Psychology*, 87(2): 268–279.

Haugh, H. and Talwar, A. (2010) "How do corporations embed sustainability across the organization?," *Academy of Management Learning & Education*, 9(3): 384–396.

Heifetz, R. (1994) *Leadership Without Easy Answers*, Harvard University Press, Cambridge, MA.

Holly, K. and Clifton, J. (2009) "It's not the economy, stupid," *BusinessWeek*, August 20.

Howell, J. and Avolio, B. (1993) "Transformational leadership, transactional leadership, locus of control, and support for innovation: key predictors of consolidated-business-unit performance," *Journal of Applied Psychology*, 78: 891–902.

Lacy, P., Arnott, J., and Lowitt, E. (2009) "The challenge of integrating sustainability into talent and organization strategies: investing in the knowledge, skills and attitudes to achieve high performance," *Corporate Governance*, 9(4): 484–494.

Lacy, P., Cooper, T., Hayward, R., and Neuberger, L. (2010) "A new era of sustainability: CEO reflections on progress to date, challenges ahead and the impact of the journey toward a sustainable economy," UN Global Compact–Accenture CEO Study, www.unglobalcompact.org/docs/news_events/8.1/UNGC_Accenture_CEO_Study_2010.pdf

Locke, E. and Latham, G. (1990) *A Theory of Goal Setting and Task Performance*, Prentice-Hall, Englewood Cliffs, NJ.

Lord, R. and Maher, K. (1991) *Leadership and Information Processing: Linking Perceptions and Performance*, Rutledge, Boston, MA.

Maslow, A. (1954) *Motivation and Personality*, Harper, New York.

Meyer, J. and Allen, N. (1991) "The three-component conceptualization of organizational commitment," *Human Resource Management Review*, 1: 61–89.

Morsing, M. and Oswald, D. (2009) "Sustainable leadership: management control systems and organizational culture in Novo Nordisk A/S," *Corporate Governance*, 9(1): 83–99.

Mowday, R., Porter, L., and Steers, R. (1982) *Employee Organization Linkages: The Psychology of Commitment, Absenteeism, and Turnover*. Academic Press, New York.

Organ, D. (1988) *Organizational Citizenship Behavior: The Good Soldier Syndrome*, Lexington Books, Lexington, MA.

Organ, D. (1990) "The motivational basis of organizational citizenship behavior," *Research in Organizational Behavior*, 12: 43–72.

Podsakoff, P., MacKenzie, S., Moorman, R., and Fetter, R. (1990) "Transformational leader behaviors and their effects on followers' trust in leader, satisfaction, and organizational citizenship behaviors," *Leadership Quarterly*, 1: 107–142.

Riggio, R., Riggio, H., Salinas, C., and Cole, E. (2003) "The role of social and emotional communication skills in leader emergence and effectiveness," *Group Dynamics: Theory, Research, and Practice*, 7(2): 83–103.

Rok, B. (2009) "Ethical context of the participative leadership model: taking people into account," *Corporate Governance*, 9(4): 461–472.

Whittington, J.L., Goodwin, V.L., and Murray, B. (2004) "Transformational leadership, goal difficulty, and task design: independent and interactive effects on employee outcomes," *Leadership Quarterly*, 15(5): 593–606.

Wofford, J.C. and Goodwin, V.L. (1994) "A cognitive interpretation of transactional and transformational leadership theories," *Leadership Quarterly*, 5(2): 161–186.

Wofford, J.C., Goodwin, V.L., and Whittington, J.L. (1998) "A field study of a cognitive approach to understanding transformational and transactional leadership," *Leadership Quarterly*, 9(1): 55–84.

Yammarino, F. and Bass, B. (1990) "Transformational leadership and multiple levels of analysis," *Human Relations*, 43: 975–995.

5 Increasing engagement through goal setting and task design

In Chapter 4 we introduced the construct of engagement, which is central to our Leading the Sustainable Organization model. In that chapter we discussed the importance of having leaders throughout the organization who utilize a full range of behaviors to engage the organization's employees in the sustainability effort. In this chapter, we extend the model by discussing two additional important antecedents to engagement: goal setting and job enrichment. Each of these motivational strategies is supported by a substantial amount of empirical research. First we discuss goal setting in the context of a comprehensive performance-planning system. Then we discuss the impact of job enrichment in engaging employees.

In addition to the well established motivational impact of goal setting and job enrichment, each is a particularly important tactic for embedding commitment to and focus on sustainability throughout an organization. This is a critical component of any successful sustainability initiative. Creating a business that is both sustainable and profitable requires efforts by people at all levels of the organization – engaging employees in the sustainability agenda is vital to success.

According to CEOs, successful efforts require managers at every level to be aware of the sustainability issues faced by the organization, to understand societal needs and concerns, to minimize risks, and to generate business opportunities (Lacy *et al.*, 2009). Moving beyond mere acknowledgement of the importance of sustainability to the point at which these issues are incorporated into day-to-day operations is a challenge. This shift can be facilitated through the incorporation of sustainability objectives into employees' performance objectives and assessments.

Goal setting and performance management

The impact of goal setting on employees' performance has been documented for a wide variety of tasks in both laboratory and field settings (Mento *et al.*, 1987; Locke and Latham, 1990). The robustness of the relationship led Mento *et al.* (1987:74) to conclude that "if there is ever to be a viable candidate from the organizational sciences for elevation to the lofty status of a scientific law of nature, then the relationships between goal difficulty, specificity/difficulty, and task performance are most worthy of serious consideration. Certainly, if nothing else, the evidence from numerous studies indicates that these variables behave lawfully." The "lawful" nature of the impact of goal setting has been described as the high-performance cycle (Locke and Latham, 1990). The process begins with a high level of challenge in the form of specific, difficult goals. When employees are committed to these goals, receive adequate feedback, and possess high self-efficacy and suitable task strategies, high performance will result. If high performance leads to desired intrinsic and extrinsic rewards, employees will experience high levels of satisfaction. High job satisfaction is, in turn, strongly related to commitment, and consequently high intentions of remaining in an organization. Employees who are satisfied and committed are then ready to accept additional challenges. Thus the cycle repeats itself. The high-performance cycle may lead to performance beyond expectations, extra-role behaviors, and commitment to the organization, as does transformational leadership.

Goal setting does not occur in a vacuum. Rather, effective goal setting takes place in the context of a comprehensive performance-management system that links performance planning and performance evaluation. This process is further enhanced by systematic process, or concurrent, feedback. Ideally, the performance-planning process should clarify for individual employees how their task performance contributes to the overall strategic goals of the organization. This links their contribution to the organization's mission and helps employees to see the significance of their contributions.

The performance-management process begins with the development of a performance plan. This involves the manager and employee reaching mutual agreement on three critical issues: the content of the job, the appropriate methods for accomplishing the job, and the expected performance outcomes of the job. Ideally, this is a two-way interaction in which both manager and employee each contribute their perspectives on these three issues, then reach a mutual understanding that becomes the basis for the ongoing performance-management process. The most important outcome of this initial agreement is the setting of performance goals that are clear, specific, and challenging. These goals should be objectively measurable and have a designated timeframe for accomplishing them.

The goals developed during the performance-planning process become the evaluative criteria that are used in the performance-evaluation process. The use of performance goals as evaluative criteria provides a clear linkage between the performance plan and the performance review. This linkage overcomes two problems normally associated with performance evaluations: contamination and deficiency. Contamination occurs when elements outside the original plan are included in the process. Deficiency occurs when important elements that should be included are omitted. By clarifying the job content and the results expected, and then using the goals developed in the performance plan as the basis for the performance evaluation, the resulting evaluation should be valid (not contaminated or deficient).

Effective performance management requires two forms of feedback. Typically, performance management focuses only on outcome feedback. Outcome feedback is formal feedback that occurs at the end of the performance-planning period during the performance-evaluation interview. However, the most effective performance-management programs also incorporate process feedback, which occurs both formally and informally between the performance-planning session and performance-evaluation interview.

A useful tool for providing process feedback is the personal management interview (PMI; Cameron, 2008). The PMI is a continuous program of recurring, one-on-one meetings between a manager and each of his or her direct employees. Underlying the PMI is the conviction that the most important aspect of an employee's experience at work is their relationship with their manager/supervisor. A person will not have a positive work experience unless their relationship with their manager has positive characteristics. Developing, improving, and nurturing the manager–employee relationship requires an intentional effort that devotes time and energy to the process. The PMI ensures that both managers and employees are committed to the process, and ensures an effective exchange of information.

When done properly, performance management allows a level of consistency and focus to be achieved throughout the organization. The mission cascades throughout the organization in such a way that each division's contribution is clearly related to the overall goals of the organization. Furthermore, divisional and departmental objectives are translated into individual employee performance plans. This is particularly important for an organization's sustainability effort, because employees want to know what their company is doing, and will take pride if it is doing well and they can understand how they contributed to this effort (Lacy *et al.*, 2010).

Goal setting and performance management can help transform a general awareness of sustainability issues into a concrete commitment that permeates an organization. This transformation usually occurs in three phases (ibid.). First, organizations must begin to measure their sustainability performance in terms of their positive and negative impact on society. Second, businesses must link their performance on sustainability to traditional business metrics and value creation (revenue growth, cost reduction, risk management, brand/reputation). Third, sustainability outcomes must be embedded within employee performance frameworks and compensation packages. Incorporating sustainability objectives into performance assessment and pay structures is particularly important, because people tend to do what they are compensated to do. This may require the development of a new set of analytics to support a company's sustainability performance management.

Another crucial element of the pay-and-performance mix is timeframe. All too often, contemporary compensation systems focus on short-term results. A "tyranny of the quarterly" mentality is not consistent with the commitments required to achieve sustainability objectives. Successful implementation of sustainability initiatives requires a reconciliation of short- and long-term metrics. Even when the right metrics are selected, tracked, and linked to value creation, a key challenge in employee remuneration relates to the time period over which performance is rewarded. The inherent tension between a reward cycle that operates on a quarterly or annual basis and the sustainability performance objectives that typically need to be assessed over a much longer timescale must be addressed directly. Doing so is essential to embedding sustainability in individual performance frameworks.

Although the link between employee engagement and productivity may be difficult to quantify in financial terms, there are many examples where performance on sustainability issues is leading to higher employee-retention rates. Better retention can, in turn, reduce the cost of recruitment and retraining, and can protect a company against the loss of corporate knowledge and experience.

We have advocated the use of goal setting and performance management as important ingredients in engaging employees in their organization's sustainability efforts. When done properly, the sustainability mission cascades throughout the organization and is represented in the performance objectives of departments and in the goals of individual employees. At this stage of the sustainability movement, this level of involvement has not yet been achieved. In our sample (Appendix A), only 33% of respondents reported that they had specific sustainability-related objectives as part of their individual performance plan. Even fewer, 24%, reported receiving regular feedback about their efforts to implement sustainability practices in their jobs.

Key principle

Sustainability goals must cascade throughout the organization and be embedded in individual employees' performance plans

---

Task design and individual job enrichment

As Ricky Griffin (1982: 153) observed, "the task that a person performs and the person to whom the individual is responsible are probably the two most basic points of contact that employees have in the organization." We discussed the importance of full-range leadership behaviors in Chapter 4, and we will return to the relationship between leaders and followers in Chapter 6. In the following section, we focus on the task that the person performs. The structure of the task has important implications for employee engagement and performance. We begin with a discussion of the job characteristics model.

The job characteristics model depicted in Figure 5.1 was developed by Hackman and Oldham (1976). Their model of task design identifies five core job dimensions: task variety, task identity, task significance, autonomy, and feedback. These are referred to as core job dimensions because they have been found to relate directly to the personal satisfaction of employees. Task variety refers to the extent to which an employee is required to develop and utilize a variety of skills and abilities to accomplish a goal. Task identity is the idea that a job involves completing an entire unit of work, rather than just a part of it. When a job has a high level of task identity, there is a sense of closure. This allows employees to see the end result of their work and the contribution they make to the organization.

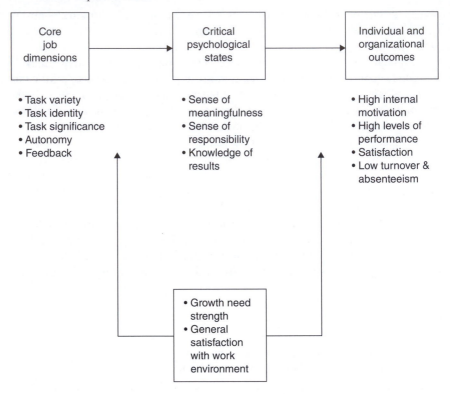

Figure 5.1 Job characteristics model

Task significance is present when an employee's efforts have a substantial and perceivable impact on other people, both within the organization and in society as a whole. When a job has significance, employees feel that their efforts really matter. Employees have job autonomy when they feel the work they accomplish is the result of personal effort and control. Feedback, in the context of the job characteristics model, refers to the extent to which the task itself provides information to the employee that they are accomplishing the desired outcome. The presence of feedback allows the outcome of the work itself to act as a reward.

Jobs that have these core dimensions are said to be enriched and have a high motivating potential (Oldham *et al.*, 1976). According to their model, the presence of these core dimensions produces three critical psychological states: a sense of meaningfulness in the work, a sense of responsibility for the work, and knowledge of the results of one's work. Task variety, task identity, and task significance combine to create a sense of meaningfulness. When these characteristic are present, employees feel that their job involves doing something that is intrinsically meaningful or worthwhile. Autonomy generates a sense of responsibility because employees feel personally responsible for a meaningful portion of their work. Feedback provides knowledge of results, which is particularly important for employees who need to have a sense of accomplishment in their work.

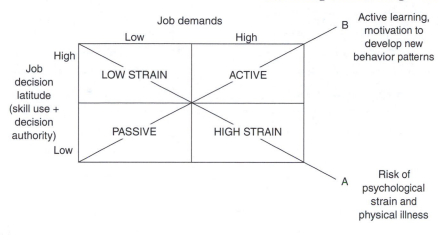

Figure 5.2 Job strain model

These critical psychological states in turn produce a variety of positive individual and organizational outcomes. Among these are high internal motivation, high quality of work performance, high satisfaction with the work, and low levels of absenteeism and turnover. Thus the impact of the core job dimensions on these outcomes is mediated by the critical psychological states.

Griffin (1991) examined the relationships between task design and job satisfaction, organizational commitment, and performance in a longitudinal field experiment. In his study, Griffin combined five task characteristics (task variety, identity, significance, autonomy, and feedback) into an overall motivating potential score (MPS). The relationships between MPS and the criteria were examined on four occasions over a four-year period. At each point, the MPS was significantly positively related to job satisfaction and organizational commitment. In the third and fourth periods, MPS also was significantly positively related to performance. It is important to note that these significant relationships were found without examining the moderating impact of employees' growth need strength and contextual satisfaction on the relationship between MPS and each criterion. This suggests that the relationship between task characteristics, as measured by MPS, and a variety of criterion variables is more direct than previously discussed. Thus enriched jobs appear to lead to a wide variety of positive outcomes, including employee job satisfaction, commitment to the organization, and performance.

More recently, Whittington *et al.* (2004) found that job enrichment as measured using the MPS was related to variety of important individual outcomes. In a field study involving employees from twelve different organizations, Whittington *et al.* found that the level of job enrichment was positively and significantly related to an employee's in-role performance, their organizational citizenship behavior, and their level of affective commitment to the organization.

The characteristic of autonomy may be particularly important for creating engagement, according to the job strain model (Karasek, 1979). According to the predicted by job strain model (Figure 5.2), the health and behavioral outcomes

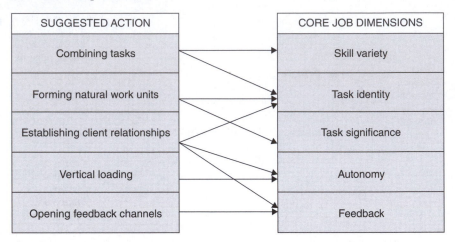

SUGGESTED ACTION	CORE JOB DIMENSIONS
Combining tasks	Skill variety
Forming natural work units	Task identity
Establishing client relationships	Task significance
Vertical loading	Autonomy
Opening feedback channels	Feedback

Figure 5.3 Guidelines for enriching a job

of job design can be focusing on the interaction of two key dimensions of the work. The first dimension is decision latitude, which combines the amount of job autonomy and the skill discretion available to the employee. The second dimension is psychological demands, which include the workload and intellectual requirements of the job.

These dimensions combine to provide four categories of jobs. A high-strain job occurs when there are high psychological demands but the employee has a low level of decision latitude. Employees in high-strain jobs are at high risk of experiencing psychological strain and stress-related illness. In contrast to the high-strain job, an active job combines high levels of psychological demands with a high level of decision latitude. While there may be moderate levels of strain in these jobs, employees in active jobs are typically motivated to learn and grow as their job evolves. A low-strain job provides a great deal of decision latitude with low demands. The final job category is a passive job, which has low demands and low levels of control.

The characteristics of an active job are consistent with the recommendations of Haugh and Talwar (2010) for embedding sustainability initiatives throughout the organization. Specifically, they advocate the need to create opportunities for employees to gain practical experience of sustainability initiatives. This increases not just their knowledge, but also their interest in and commitment to sustainability. Thus embedding sustainability should include technical and action learning opportunities.

As shown in Figure 5.3, there are several strategies available for enriching a job. Combining tasks can increase task variety and task identity. Forming natural work units around products, processes, or customers served can increase both task identity and task significance. Allowing employees to establish direct contact and develop relationships with clients improves task identity and autonomy, as well as providing important sources of feedback.

> **Key principle**
>
> Task design enhances employee engagement and performance.

Engaging the entire organization for sustainability

Goal setting in the context of a comprehensive performance plan and the redesign of jobs are important elements of engaging individual employees in an organization's sustainability effort. These are also important elements of the effort to engage the entire organization in a sustainability strategy. Benn *et al.* (2010) have identified the importance of moving away from the individual contributor mindset and moving toward a more collective perspective in successful sustainability efforts. They view sustainability as something to be shared, rather than linked to the influence of any one powerful individual. Therefore the focus of attention is shifted away from individual efforts to an insistence that the organization as a whole should take ownership of the sustainability initiatives and see that they are consistent with the organization's values.

The importance of this organization-wide effort was also echoed in Accenture's study, in which CEOs expressed the belief that a new level of sustainability cannot be reached without the broad support of the corporate culture (Lacy *et al.*, 2010). Transforming an organization's culture requires leadership from the top, as well as providing new educational opportunities and encouraging new ways of working. These new ways of working will require employees to be more aware of the inter-relations and multiple causalities within the complex environments in which their organization functions. This will also require requisite variety – adaptive systems must be in place that match the complexity of the organization's operating environment. Successful organizations will be filled with employees who have a deep understanding of how changes in the macro-environment can potentially have an impact on individuals and organizations at the micro-level.

Thus a more organic form of structuring organizations will be required. This will affect how companies connect employees with one another, and how they work together to innovate and to serve customers. Sustainable organizations must place a premium on lateral thinking and cross-functional, collaborative problem-solving. The ability to learn and store knowledge and experience, to create flexibility in problem-solving, and to balance power among interest groups can help teams improve their "adaptive capacity." The ability to change in response to the dynamic demands of the organization's operating environment will create more resilient organizations. This is particularly important as organizations begin to understand that sustainability itself is a dynamic, not static, destination.

Performance planning, as described in this chapter, helps emphasize the importance of each individual's contribution to the organization's sustainability strategy. By linking performance and compensation to these more macro-level objectives, employees are encouraged – and incentivized – to transcend the individual contributor mindset. Furthermore, by increasing the core job dimensions, in particular autonomy, jobs can be restructured in such a way that each employee can respond to the changing demands of the dynamic nature of sustainability.

Summary

- Successful implementation of sustainability strategies requires new forms of organizing and may involve the transformation of an organization's culture.
- In order for sustainability strategies to be fully embraced throughout the organization, individual performance plans and assessments must include explicit goals on sustainability.
- Performance plans and compensations systems must focus beyond short-term results (quarterly or annual) and embrace the long-term nature of sustainability.
- The successful implementation of a sustainability strategy requires a transformation from an individual contributor mindset to a more collaborative approach. The performance-management system must be leveraged to support this transformation.
- Sustainable organizations must be more flexible and responsive to the dynamic demands of the environments in which they operate. Creating these more organic organizational structures begins with redesigning individual jobs.
- While all five core job dimensions are important for engaging employees, autonomy may be the most important factor.

Discussion questions

- To what degree do you feel that the organization's key initiatives are embraced by your entire organization?
- Does your organization's performance-management system tie individual performance and compensation to organization-wide initiatives?
- Is performance planning explicitly linked to performance evaluation in your organization?
- How would you characterize your organization's culture? Is it a collection of individual contributors, or is there a genuine collaborative spirit across the organization?
- How much latitude are employees granted in making decisions in their jobs?

Key tool

Assessing your organization's performance-management system

For each of the following questions, select the answer that best describes your organization.

Table 5.1 Assessment of Performance Management System

The employees have	*Usually*	*Sometimes*	*Seldom*
1 Specific and clear goals			
2 Goals for all key areas relating to their job performance			
3 Challenging but reasonable goals (neither too hard nor too easy)			
4 The opportunity to participate in setting their goals			
5 A say in deciding how to implement their goals			
6 Deadlines for accomplishing their goals			
7 Sufficient skills and training to achieve their goals			
8 Sufficient resources (time, money, equipment) to achieve their goals			
9 Frequent feedback on how well they are progressing toward their goals			
10 Performance appraisals based on objectives set in performance-planning session			
11 Rewards (pay, promotions) allocated to them according to how well they reach their goals			
12 Goals that reflect organization-wide initiatives as well as individual performance			
13 Rewards that are tied to organizational achievement as well as individual performance			

Scoring instructions

- For each 'usually' response, assign 3 points.
- For each 'sometimes' response, assign 2 points.
- For each 'seldom' response, assign 1 point.

After assigning these points, sum the scores to create a total score.

- A total score ranging from 32–39 indicates that your organization has a solid and well integrated performance-management system.
- A total score ranging from 24–31 indicates there are some elements of an effective performance-management system in place, but additional work is needed.
- A total score below 23 indicates an underdeveloped performance-management system.

Key tool

Guidelines for personal management interviews

The PMI is an ongoing program of regular, one-on-one interviews between a manager and each of his or her direct employees.

Rationale

Underlying the PMI is the fundamental belief that the most important aspect of an employee's experience at work is their relationship with their manager/supervisor. A person will not experience positive emotion at work unless this basic relationship with their manager/supervisor has positive characteristics. The only way that this relationship can be developed, improved, and nurtured is through devoting time and energy to it. The PMI is a way to ensure managers devote enough time to their direct employees and to ensure a good two-way flow of communication.

Using the PMI

INITIAL ROLE NEGOTIATION

Prior to the regular meetings, an introductory one-off role-negotiation meeting is held. At this meeting, the manager and the direct employee discuss and negotiate the following:

- role performance;
- areas of responsibility;
- accountability and reward;
- interpersonal relationships; and
- mission, goals, and values.

Once clear agreement has been reached on the above, and any non-negotiable matters identified and justified, the decisions from the meeting are written up and form the basis for the following monthly meetings.

PRIMARY CHARACTERISTICS OF PMIs

- Regular and private.
- Major goals: continuous improvement, team-building and personal development, feedback.
- First agenda item: follow-up on action items from the previous section.
- The meeting lasts from 45 to 60 minutes.
- Major agenda items include: organizational and job issues, information and sharing, training and development, resource needs, interpersonal issues, obstacles to improvement, targets and goals, appraisal and feedback, personal issues.

- A supportive, non-punitive environment.
- Last agenda item: review of action items.

IMPLEMENTATION GUIDELINES FOR PMIs

- Regularly scheduled.
- Private.
- Free of interruptions.
- Advance preparation by both parties.
- Accountability required of both parties.
- Training of participants in advance.
- Flexibility in format.
- Action items and improvement plans.
- Feedback, praise, and development.

BENEFICIAL OUTCOMES OF PMIs

- Actually saves time.
- Institutionalizes continuous improvement.
- Improves and sustains unit effectiveness.
- Improves quality of communication.
- Maintains accountability for commitments.
- Prevents regression from off-site training.
- Provides opportunities for manager–subordinate meetings face-to-face.
- Enhances meeting effectiveness.
- Provides opportunities for training and development.
- Becomes a success experience in itself.

Key tool

Assessing your job's enrichment level

Describe your present job (or a previous paid or unpaid job you've had) using the following questionnaire. Circle the number that best describes the job. Be as objective as possible in your answers.

1. How much variety is there in your job? That is, to what extent does the job require you to do many different things at work, using a variety of your skills and talents?

    1-------------2-------------3-------------4-------------5-------------6-------------7

    Very little   Moderate   Very much

2. To what extent does your job involve doing a whole and identifiable piece of work? That is, is the job a complete piece of work that has an obvious beginning and end, or is it only a small part of the overall piece of work, which is finished by other people or by machines?

    1-------------2-------------3-------------4-------------5-------------6-------------7

    Very little   Moderate   Very much

3. In general, how significant or important is your job? That is, are the results of your work likely to significantly affect the lives or well-being of other people?

    1-------------2-------------3-------------4-------------5-------------6-------------7

    Not significant   Moderately significant   Very significant

4. How much autonomy is there in your job? That is, to what extent does your job permit you to decide on your own how to go about doing the work?

    1-------------2-------------3-------------4-------------5-------------6-------------7

    Very little   Moderate   Very much

5. To what extent does doing the job itself provide you with information about your work performance? That is, does the actual work itself provide clues about how well you are doing?

    1-------------2-------------3-------------4-------------5-------------6-------------7

    Very little   Moderate   Very much

Figure 5.4 Assessing your job's enrichment level.

Scoring instructions

To calculate the motivating potential score (MPS) for your job, add up the scores for your responses to the five questions.

*   A score of 30 or higher indicates that the job has high levels of the five core job dimensions. Thus the job has a high MPS.
*   A score in the 20–29 range indicates a moderate MPS.
*   A score of less than 20 indicates a low MPS.

References

Benn, S., Todd, L.R., and Pendleton, J. (2010) "Public relations leadership in corporate social responsibility," *Journal of Business Ethics*, 96: 403–423.

Cameron, K. (2008) *Positive Leadership: Strategies for Extraordinary Performance*, Berrett-Koehler Publishers, San Francisco, CA.

Griffin, R. (1982) *Task Design: An Integrative Approach*, Scott, Foresman and Company, Glenview, IL.

——(1991) "Effects of work redesign on employee perceptions, attitudes and behaviors: a long-term investigation," *Academy of Management Journal*, 34: 425–435.

Hackman, J. and Oldham, G. (1976) "Motivation through the design of work: test of a theory," *Organizational Behavior and Human Performance*, 16: 250–279.

Haugh, H. and Talwar, A. (2010) "How do corporations embed sustainability across the organization?," *Academy of Management Learning & Education*, 9(3): 384–396.

Karasek, R. (1979) "Job demands, job decision latitude and mental strain: implications for job redesign," *Administrative Science Quarterly*, 24: 285–308.

Lacy, P., Arnott, J., and Lowitt, E. (2009) "The challenge of integrating sustainability into talent and organization strategies: investing in the knowledge, skills and attitudes to achieve high performance," *Corporate Governance*, 9(4): 484–494.

Lacy, P., Cooper, T., Hayward, R., and Neuberger, L. (2010) "A new era of sustainability: CEO reflections on progress to date, challenges ahead and the impact of the journey toward a sustainable economy," UN Global Compact–Accenture CEO Study, www.unglobalcompact.org/docs/news_events/8.1/UNGC_Accenture_CEO_Study_2010.pdf

Locke, E. and Latham, G. (1990) *A Theory of Goal Setting and Task Performance*, Prentice-Hall, Englewood Cliffs, NJ.

Mento, A., Steel, R., and Karren, R. (1987) "A meta-analytic study of the effects of goal setting on task performance: 1966–1984," *Organizational Behavior and Human Decision Processes*, 39: 52–83.

Oldham, G., Hackman, J., and Pearce, J.(1976) "Conditions under which employees respond positively to enriched work," *Journal of Applied Psychology*, 61: 395–403.

Whittington, J.L., Goodwin, V.L., and Murray, B. (2004) "Transformational leadership, goal difficulty, and task design: independent and interactive effects on employee outcomes," *Leadership Quarterly*, 15(5): 593–606.

6 Enhancing engagement by building high-trust relationships

Introduction

> The task that a person performs and the person to whom the individual is responsible are probably the two most basic points of contact that employees have in the organization.
>
> (Griffin, 1982)

In Chapter 5 we focused on the task that a person performs. In that chapter, we discussed the role of task design and performance-management systems in creating an engaged workforce. We saw that five core job dimensions are necessary to provide a sense of meaning, responsibility, and knowledge of the outcomes of an employee's work. We also saw how clear, specific, and challenging goals could be implemented to create a high-performance cycle.

Here we turn our attention to the relationship that an employee has with his or her manager. The importance of this relationship cannot be emphasized enough. This is reinforced by Buckingham and Coffman (1999), who observed that "employees don't quit companies, they quit managers." In a more forceful tone, Bedeian and Armenakis (1998: 59) state that "As assuredly as Gresham's Law states that bad money drives out good money, incompetent managers, wherever situated, inevitably drive away good employees."

The Society for Human Resource Management affirms the importance of the quality of the leader–follower relationship, stating that the number one factor that influences employee commitment is the manager–employee relationship (Lockwood, 2007). The manager creates the connection between the employee and the organization, and as a result, the manager–employee relationship is often the "deal-breaker" in relation to retention. Employees who trust their managers appear to have more pride in the organization and are more likely to feel they are applying their individual talents for their own success and that of the organization. Yet only a small majority of employees feel their manager has good knowledge of what they do and promotes the use of their unique talents (ibid.).

In this chapter, we utilize the framework of leader–member exchange (LMX) theory to examine of the quality of the leader–follower relationship. This relationship provides the immediate context for the emergence of trust, which is the foundation for engaging employees in the purpose and strategies of the organization.

Leader–member exchange: the quality of the relationship between leaders and employees

In his seminal book *Leadership*, Burns (1978) stated that leadership is a process that takes place in the context of a relationship between leader and follower. No theory of leadership addresses the relational nature of leadership more directly than LMX theory (Dienesch and Liden, 1986; Liden *et al.*, 1993). The basic premise of LMX theory is that leaders develop different types of exchange relationships with their subordinates as the relationship evolves. According to LMX theory, leaders differentiate among their followers. Rather than using the same style with all followers, leaders develop different styles for members of in-groups and out-groups. In-group members are likely to receive assignments to interesting and desirable tasks, have greater responsibility and authority delegated to them, have more information shared with them, participate in making some of the leader's decisions, and receive personal support and approval (Yukl, 2009). In-group members are included in the inner life of the organization, while out-group members are excluded (Sparrowe and Liden, 1997).

Relationships between the leader and members of the in-group are characterized by mutual trust, respect, liking, and reciprocal influence. Followers in these high-quality LMX relationships have frequent interactions with their leaders and enjoy the support, confidence, and consideration of the leader (Howell and Hall-Merenda, 1999). Because in-group members receive a higher level of individualized consideration than out-group members, it is expected that in-group followers will perceive leaders as more transformational than out-group members (Graen and Uhl-Bien, 1995). Conversely, relationships with out-group members are characterized by downward influence and role-defined relations. These low-quality LMX relationships are based on the formal roles and exchange relationships that characterize transactional leadership (Burns, 1978; Bass and Avolio, 1994).

Boyd and Taylor (1998), who examined the role of friendship in the leader–follower relationship, have investigated a similar line of thought. Following Graen and Uhl-Bien (1995), they argue that the nature of the relationship evolves over time and affects a variety of organizational outcomes, including communication, mutual disclosure, and support. According to Boyd and Taylor (1998), the leader–follower relationship develops through at least four stages. In the initial, LMX potential stage, the relationship between leader and follower is influenced by similarity and attraction evaluations on the part of both parties. During the second, exploration/low-LMX stage, relationships are superficial and lack the breadth of activity and depth of intimacy that characterize more mature LMX relationships. There is a lack of openness and rich interaction, thus leader–follower relationships are characterized by caution and tentativeness. As the relationship evolves into stage three, the leader–follower relationship becomes more friendly, relaxed, and casual. During this phase, both members of the relationship are more willing to evaluate and be evaluated, test each other concerning minimal expectations and limits of influence, and begin to define the limits of mutual trust.

Stage four represents a high-LMX or mature stage of the leader–follower relationship, where a high degree of mutual trust, respect, loyalty, intimacy, openness, honesty, and obligation is present (Graen and Uhl-Bien, 1995). In this phase, the leader–follower relationship would likely involve mutual influence and reciprocity, representing a shift in leader behavior from transactional to transformational leadership (Boyd and Taylor, 1998).

Contemporary LMX theory focuses on how the quality of the leader–follower relationship affects a wide variety of outcomes. These outcomes include employee performance, overall job satisfaction, satisfaction with supervision, organizational commitment, turnover intentions, perceptions of fairness, trust in the supervisor, and innovative behavior of followers (Graen and Uhl-Bien, 1995; Gerstner and Day, 1997; Scott and Bruce, 1998). Recent LMX research examines the dyadic and organizational aspects that enhance LMX, including effective communications (Yrle *et al.*, 2003; Stringer, 2006); organizational justice (Bhal, 2006; Burton *et al.*, 2008); perceived organizational support of supervisors (Erdogan and Enders, 2007; Sluss *et al.*, 2008); and the emotional intelligence of supervisors and subordinates (Sears and Holmvall, 2010). Additional studies have also found positive relationships between LMX and employee trust in supervisors (Flaherty and Pappas, 2000); employee turnover (Morrow *et al.*, 2005); and job satisfaction (Stringer, 2006). It is clear that the quality of the leader–follower relationship has important implications for engaging employees.

Key principle

High-quality relationships between leaders and their employees are essential to creating an engaged workforce.

Significance of trust in the leader–follower relationship

High-quality relationships provide the fertile soil necessary to create trust, which has emerged as a critical element in creating sustainable, high-engagement organizations. Clear implications for the effect of trust in leadership on follower behavior have been emphasized in publications in the popular management press (Kouzes and Posner, 2008; Covey, 1990, 2006; Galford and Drapeau, 2002, 2003) and in scholarly research articles (Colquitt *et al.*, 2007; Mulder, 2009). Trust is not only important for sustaining individual and organizational effectiveness (McAllister, 1995), but it also lies at the heart of relationships and influences the behavior of each party toward the other (Robinson, 1996). The leader–follower relationship is no exception. When subordinates trust a leader, they are willing to be exposed to the leader's actions, and are certain that their interests will not be abused (Mayer *et al.*, 1995). If this trust is broken, it can have severe undesirable effects (Dirks and Ferrin, 2002).

In the process of motivating followers to implement their shared vision, transformational leaders become role models for their followers, demonstrating what it means to persevere and make self-sacrifices when needed (Jung and Avolio, 2000). Through observation, followers develop trust in their leaders because of their personal commitment to achieving the vision. Furthermore, transformational leaders empower and encourage followers to think for themselves, which instills trust in the leader (Avolio and Bass, 1995). On the flip-side, transformational leadership can involve moving followers from the familiar to the unfamiliar. Followers may experience higher levels of fear, anxiety, frustration, and uncertainty; all of which can be alleviated by the trust they have in their leaders (Kotter, 1996).

Galford and Drapeau (2002) identify three kinds of trust in organizations. Strategic trust is created when employees and other stakeholders believe that the organization is pursuing the right goals and strategies. Organizational trust is created when people believe that the operational processes and decision-making are being conducted in the right way. Finally, interpersonal trust occurs when employees trust their leaders and this trust is reciprocated by the leaders toward the employees.

Interpersonal trust can be further defined as expectancy held by an individual that another individual can be relied upon (Rotter, 1967). There are two types of interpersonal trust – cognition-based trust and affective-based trust (McAllister, 1995). Cognition-based trust comes from knowledge of an individual that provides evidence of trustworthiness. Affective-based trust comes from the emotional bonds between individuals. Followers of transformational leaders are likely to have both types of trust in their leaders because of the role-modeling they have observed in their leaders and the interpersonal ties that develop between them.

Key principle

Trust is vital for creating and sustaining an engaged workforce.

Covey (2006) states that, at the most fundamental level, trust is really all about credibility. Credibility has two dimensions: character and competence.

The character dimension of trust

The character dimension is based on integrity and intent.

Integrity

Integrity refers to honesty – yet it involves more than honesty. The root word for integrity is "integer," which refers to being whole. In the context of leader–follower relationships, this means there is congruence between the leader's talk and his or her walk. Leaders who exhibit integrity have the courage to stick to their values, even when it may involve a high personal cost.

Integrity is an ever-increasing component of a leader's character. Leaders who strive for integrity are always willing to confront their own lack of integrity. Leaders can increase in integrity by making and keeping commitments to themselves. These self-commitments are foundational for making and keeping commitments to others (Covey, 2006). Those who lead with integrity also have a core set of values that are non-negotiable. In other words, these leaders know who they are and what they stand for. Although those who lead with integrity do so from a strong foundation of core values, they remain open to the opinions, fears, concerns, and inputs of those around them. This requires both humility and courage (ibid.). The humble leader is willing to admit that there may be important pieces of information of which they are not aware. Courage enters into this when the leader makes a discovery and is willing to act on it.

Intent

The second element of character is intent. Intent refers to motives and the agendas that underlie our behaviors. While motives cannot be directly observed, trust grows when people perceive that a leader is being straightforward and acting in ways that will be mutually beneficial for both leader and follower. Indeed, it is the follower's perceptions of the leader's motives that are crucial.

The importance of the follower's perceptions of a leader's behavior has been emphasized by Lord and co-workers (Lord and Maher, 1993; Hall and Lord, 1995). According to their perspective, leadership is not located solely in the leader or the follower, rather it involves the behaviors, traits, and outcomes produced, as interpreted by the followers (Lord and Maher, 1993). In fact, Lord and Maher (1993) define leadership as the process of being perceived as a leader. This perspective is also supported by Yammarino and Dubinsky (1994), who argue that the effects of leadership are the result of subordinates' perceptions of the leader's behavior, not the leader's behavior *per se*. According to Yammarino and Dubinsky (1994) and Avolio and Yammarino (1990: 193), transformational leadership is thus "in the eyes of the beholder."

The importance of intent has been emphasized by Kanungo and Mendonça (1996), who identified two contrasting leadership motive patterns that provide an answer to this question: altruistic and egotistic. The altruistic motive pattern is rooted in the intent to benefit others. It is this motive of genuine caring that inspires the greatest level of trust (Covey, 2006). Conversely, the egotistic pattern is based in the intent to benefit self. Kanungo and Mendonça (1996) differentiate further between the two motive patterns by looking at the operative needs and influence strategies of leaders who operate from each of the motive patterns. The operative needs dimension refers to the leader's combination of needs for affiliation, power, and achievement (Boyzatis, 1973, 1982; McClelland and Burnham, 1995). Influence strategy refers to the power bases (French and Raven, 1959) and influence tactics used by the leader.

According to Kanungo and Mendonça (1996), leaders who operate from the egotistic motive pattern are driven by avoidance affiliation, personal power, and personal achievement. Individuals whose relationships are characterized by avoidance affiliation use relationships in order to protect themselves. Thus the need for affiliation is based on the individual's sense of insecurity and manifests itself in "non-interfering" and "easy-to-get-along" behaviors, even when the job situation demands otherwise. These leaders are reluctant to give negative feedback to subordinates. They yield to employee requests because they do not want to incur the employee's displeasure, and they do so without regard to the effect of their behavior on the need for equity, due process, and order in the workplace. Consequently, followers are left in a position of weakness without a sense of what might happen next. Followers do not know where they stand in their relationship with their manager, or even what they ought to be doing (McClelland and Burnham, 1995).

In contrast to the avoidance affiliation of the egotistic leader, altruistic leaders have an affiliative interest in their followers. These leaders are motivated primarily by a genuine interest in others, and emphasize relationships in a manner that is consistent with the demands of the job. These leaders relate to followers as individuals with ideas and resources. These relationships are characterized by a desire on the part of the leader to create a mutually beneficial agenda. Thus followers are viewed as partners in the problem-solving and related activities necessary for attaining organizational objectives. Consequently, supportive feelings permeate the interpersonal relationships between altruistic leaders and their followers.

Egotistic leaders have a high need for personal power. They are preoccupied with their own interests and concerns. This self-interest is often pursued even at the cost of the organization's welfare and effectiveness. These leaders demand and expect followers' loyalty and efforts to be directed toward the achievement of the leader's personal goals. Because of this, these leaders tend to draw on the resources inherent in the power base of their position. The personal power need of egotistic leaders seems to be rooted in a deep-seated sense of insecurity, which manifests itself in dictatorial forms of behavior. This behavior, in turn, leads to defensive relations with their followers. Insensitive to the needs of their followers, they expect unquestioning obedience to, and compliance with, their authority and decisions (Howell and Avolio, 1992).

Altruistic leaders are characterized by an institutional need for power. In contrast to the individualized need for power, these leaders have a dominant preoccupation with the concerns, goals, and interests of the organization and its members. They yield their self-interest to that of the organization. They draw primarily on the resources of their personal power base – that is, their expertise or attraction as perceived by their followers. While relying primarily on personal power bases, these leaders also may rely on rewards and sanctions as means of control and influence. However, they do so impartially and equitably. In contrast to the personal power need of egotistic leaders, the institutional power need is derived from identification with, and commitment to, the organization's objectives and interests. Thus power is a tool to serve the needs of the organization and its members. It is manifested in behaviors and feelings that serve to help and support followers in accomplishing their tasks. Leaders who operate from an institutional power need establish open communication with their followers and create a climate in which followers are encouraged to provide suggestions and criticisms of the leader's decisions and actions (ibid.).

Typically, individuals high on the achievement motive derive satisfaction from achieving their goals. They tend to pursue achievement almost as an end in itself. While assuming a high degree of personal responsibility, they also tend to be self-oriented by viewing organizational resources and support primarily in terms of their own objectives. These individuals may be motivated by either personal achievement or social achievement. Egotistic leaders are driven by personal achievement motives, and are more likely to engage in behaviors that benefit self rather than others. In fact, because they focus on personal improvement and doing things better by themselves, they want to do things themselves (McClelland and Burnham, 1995), and have difficulty relinquishing control to others through delegation. In contrast, altruistic leaders are driven by a social achievement motive. These leaders show a concern for others, and initiate efforts that focus on individual and collective capability. They are concerned with creating a better quality of life, and seek to engage in meaningful organizational and social action in order to influence the common good (Mehta, 1994; Kanungo and Mendonça, 1996).

Egotistic leaders also differ from altruistic leaders with regard to the influence strategies they employ. According to Kanungo and Mendonça (1996), egotistic leaders seek to control followers' behavior by using the legitimate rights of their position to coerce followers into compliance or by manipulating rewards. In contrast, altruistic leaders seek to empower followers and operate from a personal power base of expertise and attraction.

Covey (2006) provides several guidelines for improving the intent dimension of a leader's character. First, those who wish to lead with integrity must constantly examine and refine their motives. This should become a personal discipline; however, the most admired leaders also have a close circle of trusted confidants who they allow, and even encourage, to challenge their motives. Leading with integrity also involves a level of transparency. This means that the leader declares his or her intent, eliminating speculation about hidden agendas. Finally, Covey suggests that leaders choose to operate from a mindset of abundance. This eliminates the zero-sum games and win–lose perspectives that are typical of many relationships. The abundance mindset assumes there is enough room for everybody to receive benefits and thus eliminates dysfunctional competition.

Key principle:

Character is the foundation upon which trust is built. Character involves both integrity and intent.

The competence dimension of trust

The competence dimension is based on capabilities and results.

Capabilities

The importance of character in building trust is unquestioned. Yet if a leader has integrity and operates from an altruistic motive pattern, but is not competent, people will not be willing to follow. Capabilities include the aptitudes that allow people to perform with excellence. These aptitudes include natural gifts and abilities, as well as attitudes. Capabilities also include skills and knowledge. Skills are the things a leader does well. Knowledge encompasses learning, under-standing, and insight. Finally, capabilities also involve a leader's style and unique personality.

Results

Results are the outcomes that are expected from employee effort. These usually include the bottom-line goals of profitability, market share, and returns on investment. While these are important, we believe that other, "softer" results are just as important. These softer results include employee satisfaction and commitment, as well as employee growth and retention.

Leaders can increase their competency by first becoming aware of their unique competencies. Too many leaders confuse responsibility with competency (Stanley, 2006). While they may be responsible for many functional areas, they need to lead from their strengths. Failure to do so dilutes a leader's effectiveness.

Competent leaders also focus on results, not activities. These leaders do not confuse hard work, even by well intentioned employees, with meeting actual performance expectations. When performance expectations are met, these leaders share the praise with everyone who contributed. Yet when performance falls short of expectations, they look to themselves first to determine how their own leadership, or lack thereof, contributed to the shortfall. By doing this, trusted leaders practice the "window and mirror" principle advocated by Collins (2001).

Key principle

The development of trust also requires competence. Competence involves both capabilities and consistent results.

Trust: a benefit–cost analysis

When followers perceive that their leader is operating from an altruistic motive pattern, it is easy to trust that leader. The benefits from trust in the leader–follower relationship are significant, and trusted leaders have a potential advantage over leaders who are not trusted by their followers (Covey, 1990; Galford and Drapeau, 2002, 2003). Yet when leaders are operating from an egotistic motive pattern, it is difficult, if not impossible, for trust to develop. Covey (2006) states that when trust is absent, relationships and organizations pay a "trust tax" due to a lack of candor, hidden agendas, and dysfunctional organizational politics.

When leaders are perceived to be operating from an altruistic motive pattern, high levels of trust are likely to be present in the relationship. This creates a "dividend" that accrues in the form of lower costs and higher transaction speed (Covey, 2006). Galford and Drapeau (2003) have expanded Covey's idea of a trust dividend, identifying several additional benefits that are present when employees have high levels of trust. High levels of trust eliminate the costs associated with political maneuvering. This allows employees the autonomy to make decisions based on the need of the organization without having to look over their shoulder. The level of freedom provided by this trust fuels employees' passion about the organization and unleashes their creative potential. Thus trust fosters an environment of innovation that is critical in creating a sustainable organization.

Trust is reciprocal and contagious. People tend to trust those who have first trusted them. As trust between individuals begins to grow, a climate of trust begins to permeate the entire organization. This internal trust eventually spreads to external stakeholders as well. Thus the organization enjoys a tremendous reputation in the marketplace. This leads to an important contribution to the engagement factor. The external reputation as a trusted organization facilitates the attraction of quality employees, and the internal climate of trust enhances the retention of those quality employees. The following list presents several benefits of trust (Galford and Drapeau, 2002; Covey, 2006):

- lower operating costs;
- faster transactions;
- greater employee decision-making latitude;
- greater passion;
- increased focus;
- heightened levels of innovation;
- time to make the right decision;
- a "trust contagion" – trust breeds trust;
- enhanced recruitment of quality employees;
- enhanced retention of quality employees;
- improved performance.

Enemies of trust

The benefits associated with high trust are clear. Yet many organizations are permeated by what Galford and Drapeau (2002, 2003) call the "enemies of trust." These enemies can be sorted into three broad categories: inadequate communication, misbehavior, and situations not remedied or addressed. The enemies of trust identified by Galford and Drapeau (2002) are summarized below.

Inadequate communication

- People whose personal agendas are at odds with the organization's strategy.
- People with volatile personalities.
- Rescinding of an agreement.
- People whose behavior reflects controlled vengeance.
- Inconsistent messaging.
- Lack of congruence between espoused values and actual leader behavior.

Misbehavior

- People whose needs for promotion, power, and recognition are potentially lethal to the organization.
- The unintended consequences of a particular event, announcement, or initiative.
- The constancy of corporate reorganizations or management changes.
- Rapidly changing situations.
- Benevolence in the wrong place.
- Inadequate amounts of feedback, or false feedback.
- Ignoring rumors, unpleasant news and "undiscussable issues."
- Inconsistent enforcement of performance standards and policies.
- Excessively strict or inflexible standards.

Situations not remedied or addressed

- A culture of blame, rather than accepting responsibility for errors.
- Removal of everyday perks without proper explanation.
- Paralysis in the face of difficulty.
- Incompetence.
- Failing to trust others.

Relationship between full-range leadership and trust

In addition to the dividends associated with trust, when followers trust their leaders they exhibit more organizational citizenship behaviors (OCBs) that better equip the leader to accomplish the goals of the organization (McAllister, 1995; Colquitt *et al.*, 2007). Followers' trust in the leader may boost their confidence in the character of the leader, thus encouraging them to reciprocate with care and concern for their leaders (Dirks and Ferrin, 2002). When followers trust their leaders, they perform better and exhibit less counterproductive behavior that may come from their intentions to quit (Colquitt *et al.*, 2007).

Several studies have examined trust as an outcome of transformational leadership, among other individual and organizational outcomes. Podsakoff *et al.* (1990) found that intellectual stimulation was negatively associated with a measure of trust that assessed how fairly followers felt they were treated by their managers. Because intellectual stimulation involves challenging the assumptions that support the *status quo*, they suggested that this result may be due to the association between intellectual stimulation and higher levels of role ambiguity, conflict, and stress in the workplace; but this relationship may be unique to the short term, becoming positive in the long term. Relationships between other dimensions of transformational leadership and trust were positive. Gillespie and Mann (2004) found that all components of transformational leadership were positively correlated with trust, although the correlation for intellectual stimulation was lower than the others.

Podsakoff *et al.* (1996) assessed trust by asking followers how fairly they felt they were treated by their managers. They found that when leaders provided an appropriate model, individualized support, and fostered acceptance of group goals (all aspects of transformational leadership), employee trust was higher. Podsakoff *et al.* also found, however, that trust was associated with the greatest number of moderating effects (using the substitutes for leadership model; Kerr and Jermier, 1978), more than other outcomes of transformational leadership.

In a meta-analysis to summarize and evaluate primary relationships between trust in leadership and twenty-three constructs, Dirks and Ferrin (2002) found a strong correlation between transformational leadership and trust. They suggested that the distinction between transformational leadership and trust is therefore unclear and should be examined further, particularly focusing on measurement issues and on causal processes involved. However, Dirks and Ferrin also found that trust was strongly related to attitudes (e.g. job satisfaction and organizational commitment), followed by citizenship behaviors, and finally job performance.

These results are similar to those obtained for transformational leadership and attitudes [satisfaction with the leader (Podsakoff *et al.*, 1990; Bycio *et al.*, 1995) and employees' affective commitment to the organization (Bycio *et al.*, 1995; Whittington *et al.*, 2004)]; OCBs (Whittington *et al.*, 2004); and job performance (Bass, 1985; Hater and Bass, 1988; Yammarino and Bass, 1990a; 1990b; Keller, 1992; Howell and Avolio, 1993; Bass and Avolio, 1994; Whittington *et al.*, 2004).

The strong connection between full-range leadership and trust has been developed in detail by Avolio (2011). He identifies four "core principles" that form the logic for the development of trust in leaders and its implications. According to Avolio, we place our highest trust in those with whom we identify and those who are often vulnerable. First, exemplary leaders are willing to be vulnerable and transparent with their followers. They trust their followers with their egos (Lencioni, 1998) and the employees embrace and understand these vulnerabilities. Leaders who are willing to be vulnerable demonstrate the ability to empathize with others and fully appreciate how others are feeling and reacting to the situations they are facing. When employees embrace a leader's vulnerabilities, a sense of identification between the leader and follower is created. This sense of identification creates a sense of affective commitment (Allen and Meyer, 1990) that transcends mere compliance with policies and requests.

According to the Leading the Sustainable Organization model (Figure1.1), trust acts as a moderator of the relationships between full-range leadership, task design, goal setting and engagement. Trust also acts as a moderator of the relationships between engagement and both in-role performance and OCBs. As a moderator in this model, trust acts to enhance these relationships. In other words, trust makes the connections between the antecedents of engagement stronger than they would be if trust was not present. Additionally, trust also enhances the relationship between engagement and its consequences: in-role and extra-role performance (OCBs).

Key principle

Trust is a critical element in the creation of a sustainable organization because it enhances the relationships between the antecedents and consequences of engagement.

Summary

- The creation of a sustainable organization requires an engaged workforce.
- Engagement is created when leaders utilize the full range of leadership behaviors, enrich jobs by building in the core job dimensions, and clarify expectations through goal setting and consistent performance-management practices. These are antecedents to engagement.
- The relationship between these antecedents and engagement is enhanced by high levels of trust between leaders and their followers.
- When employees are engaged, they exhibit high levels of both in-role and extra-role performance in the form of OCBs. These are the consequences of engagement.
- The relationship between engagement and its consequences is enhanced by high levels of trust between leaders and their followers.
- Trust emerges in high-quality (high LMX) relationships between leaders and followers.
- Trust is a function of both character and competence.
- Trusted leaders have integrity and operate from an altruistic motive pattern that is perceived as intent to create mutual benefits.
- Trusted leaders demonstrate capabilities and deliver consistent results over time.
- The absence of trust "taxes" the organization, while the presence of trust creates the "dividends" necessary to create a sustainable organization.

Discussion questions

- Do the employees in your organization believe that the organization is pursuing the right goals and strategies? (Strategic trust.)
- Do the employees believe that organizational processes and decision-making are fair? (Organizational trust.)

- Is trust reciprocated in your organization? Do leaders trust their employees and do employees trust their leaders? (Interpersonal trust.)
- In general, is your organization prone to in-groups and out-groups?
- How does the presence of in- and out-groups affect the climate of trust in your organization?
- What trust "taxes" have you observed in your organization?
- Identify the specific trust "dividends" you have experienced in your organization.

Key tool

Table 6.1 LMX scale: evaluating the quality of your relationship with your leader

	Strongly disagree	Disagree	Slightly disagree	Neutral	Slightly agree	Agree	Strongly agree
1 Regardless of how much power he/she has built into his/her position, my supervisor would be personally inclined to use his/her power to help me solve problems in my work	1	2	3	4	5	6	7
2 I can count on my supervisor to "bail me out," even at his/her own expense, when I really need it	1	2	3	4	5	6	7
3 My supervisor understands my problems and needs	1	2	3	4	5	6	7
4 My supervisor recognizes my potential	1	2	3	4	5	6	7
5 My supervisor has enough confidence in me that he/she would defend my decisions if I were not present to do so	1	2	3	4	5	6	7
6 I usually know where I stand with my supervisor	1	2	3	4	5	6	7

Scoring instructions

- To evaluate the quality of your relationship with your manager, sum the responses to the questions.
- A total score of 36–42 indicates a very high-quality relationship.
- A total score between 28 and 35 indicates a quality relationship, but one that has room for additional improvement.
- Scores below 28 indicate a low-quality relationship.

Key tool

Table 6.2 Trust scale: evaluating the trust you have in your leader

	Strongly disagree	Disagree	Slightly disagree	Neutral	Slightly agree	Agree	Strongly agree
1 I place a great deal of trust in my manager	1	2	3	4	5	6	7
2 I am willing to rely on my manager	1	2	3	4	5	6	7
3 I feel quite confident that my leader will always try to treat me fairly	1	2	3	4	5	6	7
4 My managers would never try to gain an advantage by deceiving workers	1	2	3	4	5	6	7
5 I have complete faith in the integrity of my manager	1	2	3	4	5	6	7
6 I feel a strong loyalty to my leader	1	2	3	4	5	6	7
7 I would support my leader in almost any emergency	1	2	3	4	5	6	7
8 I have a divided sense of loyalty toward my leaders	1	2	3	4	5	6	7

Scoring instructions

* To evaluate the quality of your relationship with your manager, sum the responses to the questions.
* A total score of 48–56 indicates an extremely high level of trust.
* A total score between 32 and 47 indicates a high level of trust in the relationship; however, opportunities to improve on the level of trust still exist.
* Scores below 32 indicate there is a low level of trust in this relationship.

References

Allen, N. and Meyer, J. (1990) "The measurement and antecedents of affective, continuance and normative commitment to the organization," *Journal of Occupational Psychology*, 63: 1–18.

Avolio, B. (2011) *Full Range Leadership Development*, Sage, Thousand Oaks, CA.

Avolio, B. and Bass, B. (1995) "Individualized consideration is more than consideration for the individual when viewed from multiple levels of analysis," *Leadership Quarterly*, 6(2): 199–218.

Avolio, B. and Yammarino, F. (1990) "Operationalizing charismatic leadership using a level of analysis framework," *Leadership Quarterly*, 1: 193–208.

Bass, B. (1985) *Leadership and Performance Beyond Expectations*, The Free Press, New York.

Bass, B. and Avolio, B. (1994) *Improving Organizational Effectiveness through Transformational Leadership*, Sage, Thousand Oaks, CA.

Bedeian, A. and Armenakis, A. (1998) "The cesspool syndrome: how dreck floats to the top of declining organizations," *Academy of Management Executive*, 12(1): 58–67.

Bhal, K.T. (2006) "LMX–citizenship behavior relationship: justice as a mediator," *Leadership & Organization Development Journal*, 27(1/2): 106–117.

Boyd, N. and Taylor, R. (1998) "A developmental approach to the examination of friendship in leader follower relationships," *Leadership Quarterly*, 9: 1–25.

Boyzatis, R. (1973) "Affiliation motivation," in D. McClelland (ed.), *Human Motivation: A Book of Readings*, General Learning Press, Morristown, NJ, pp. 252–278.

——(1982) *The Competent Manager: A Model for Effective Performance*, John Wiley, New York.

Buckingham, M., and Coffman, C. (1999) *First, Break All the Rules: What the World's Greatest Managers Do Differently*, Simon & Shuster, New York.

Burns, J.M. (1978) *Leadership*, Harper & Row, New York.

Burton, J., Sablynski, C., and Sekiguchi, T. (2008) "Linking justice, performance, and citizenship via leader–member exchange," *Journal of Business and Psychology*, 23(1–2): 51–61.

Bycio, P., Hackett, R., and Allen, J. (1995) "Further assessments of Bass's (1985) conceptualization of transactional and transformational leadership," *Journal of Applied Psychology*, 80: 468–478.

Collins, J. (2001) *Good to Great: Why Some Companies Make the Leap ... and Others Don't*, Harper Business, New York.

Colquitt, J.A., Scott, B.A., and LePine, J.A. (2007) "Trust, trustworthiness, and trust propensity: a meta-analytic test of their unique relationships with risk taking and job performance," *Journal of Applied Psychology*, 92: 909–927.

Covey, S. (1990) *Principle–Centered Leadership*, Simon & Schuster, New York.

Covey, S.M.R. (2006) *The Speed of Trust*, Free Press, New York.

Dienesch, R.M. and Liden, R.C. (1986) "Leader–member exchange model of leadership: a critique and further development," *Academy of Management Review*, 11(3): 618–634.

Dirks, K.T. and Ferrin, D.L. (2002) "Trust in leadership: meta-analytic findings and implication for research and practice," *Journal of Applied Psychology*, 87: 611–628.

Erdogan, B. and Enders, J. (2007) "Support from the top: supervisors' perceived organizational support as a moderator of leader–member exchange to satisfaction and performance relationships," *Journal of Applied Psychology*, 92(2): 321–330.

Flaherty, K.E. and Pappas, J.M. 2000. 'The role of trust in sales person–sales manager relationships," *Journal of Personal Selling & Sales Management*, 20(4): 271–278.

French, J. and Raven, B. (1959) "The bases of social power," in D.P. Cartwright (ed.) *Studies in Social Power*, Institute for Social Research, Ann Arbor, MI, pp. 150–167.

Galford, R. and Drapeau, A. (2002) *The Trusted Leader*, The Free Press, New York.

——(2003) "The enemies of trust," *Harvard Business Review*, 81(2): 88–95.

Gerstner, C.R. and Day, D.V. (1997) "Meta-analytic review of leader–member exchange theory: correlates and construct issues," *Journal of Applied Psychology*, 82(6): 827–844.

Gillespie, N.A. and Mann, L. (2004) "Transformational leadership and shared values: the building blocks of trust," *Journal of Managerial Psychology*, 19: 588–607.

Graen, G.B. and Uhl-Bien, M. (1995) "Relationship-based approach to leadership: development of leader–member exchange (LMX) theory of leadership over 25 years: applying a multi-level multi-domain perspective," *Leadership Quarterly*, 6(2): 219–247.

Griffin, R.W. (1982) *Task Design: An Integrative Approach*, Scott-Foresman, Glenview, IL.

Hall, R. and Lord, R. (1995) "Multi-level information-processing explanations of followers' leadership perceptions," *Leadership Quarterly*, 6: 265–287.

Hater, J. and Bass, B. (1988) "Superiors' evaluations and subordinates' perceptions' of transformational and transactional leadership," *Journal of Applied Psychology*, 73: 695–702.

Howell, J. and Avolio, B. (1992) "The ethics of charismatic leadership: submission or liberation?," *Academy of Management Executive*, 6(2): 43–54.

Howell, J. and Avolio, B. (1993) "Transformational leadership, transactional leadership, locus of control, and support for innovation: key predictors of consolidated-business-unit performance," *Journal of Applied Psychology*, 78: 891–902.

Howell, J. and Hall-Merenda (1999) "The ties that bind: the impact of leader–member exchange, transformational and transactional leadership, and distance on predicting follower performance," *Journal of Applied Psychology*, 84: 680–694.

Jung, D.I. and Avolio, B.J. (2000) "Opening the black box: an experimental investigation of the mediating effects of trust and value congruence on transformational and transactional leadership," *Journal of Organizational Behavior*, 2: 949–964.

Kanungo, R.N. and Mendonça, M. (1996) *Ethical Dimensions of Leadership*, Sage, Thousand Oaks, CA.

Keller, R. (1992) "Transformational leadership and the performance of research and development project groups," *Journal of Management*, 18: 489–501.

Kerr, S. and Jermier, J. (1978) "Substitutes for leadership: their meaning and measurement," *Organizational Behavior and Human Performance*, 22: 375–403.

Kotter, J.P. (1996) *Leading Change*, Harvard Business School Press, Boston, MA.

Kouzes, J. and Posner, B. (2008) *The Leadership Challenge: How to Get Extraordinary Things Done in Organizations*, 4th edn, Josey-Bass, San Francisco, CA.

Lencioni, P. (1998) *The Five Temptations of a CEO*, Josey-Bass, San Francisco, CA.

Liden, R.C., Wayne, S.J., and Stilwell, D. (1993) "A longitudinal study on the early development of leader–member exchanges," *Journal of Applied Psychology*, 78(4): 662–674.

Lockwood, N. (2007) "Leveraging employee engagement for competitive advantage," *SHRM Research Quarterly*. www.improvedexperience.com/doc/02_Leveraging_Employee_Engagement_for_Competitive_Advantage2.pdf

Lord, R. and Maher, K. (1993) *Leadership and Information Processing: Linking Perceptions and Performance*, Rutledge, Boston.

Mayer, R., Davis, J., and Schoorman, F.D. (1995) "An integrative model of organizational trust," *Academy of Management Review*, 20: 709–734.

McAllister, D. (1995) "Affect- and cognition-based trust as foundations for interpersonal cooperation in organizations," *Academy of Management Journal*, 1: 24–59.

McClelland, D. and Burnham, D. (1995) "Power is the great motivator," *Harvard Business Review*, 73: 126–139.

Mehta, P. (1994) "Empowering the people for social achievement," in R. Kanungo and M. Mendonça (eds) *Work Motivation: Models for Developing Countries*, Sage, New Delhi, pp. 161–183.

Morrow, P.C., Suzuki, Y., Crum, M.R., Ruben, R., and Pautsch, G.(2005) "The role of leader–member exchange in high turnover work environments," *Journal of Managerial Psychology*, 20(8): 681–694.

Mulder, L. (2009) "Sanctions and moral judgments: the moderating effect of sanction severity and trust in authorities," *European Journal of Social Psychology*, 39: 255–269.

Podsakoff, P., MacKenzie, S., Moorman, R., and Fetter, R. (1990) "Transformational leader behaviors and their effects on followers' trust in leader, satisfaction, and organizational citizenship behaviors," *Leadership Quarterly*, 1: 107–142.

Podsakoff, P., MacKenzie, R., and Bommer, W. (1996) "Transformational leader behaviors and substitutes for leadership as determinants of employee satisfaction, commitment, trust, and organizational citizenship behaviors," *Journal of Management*, 22: 259–298.

Robinson, S.L. (1996) "Trust and breach of the psychological contract," *Administrative Science Quarterly*, 40: 574–598.

Rotter, J. (1967) "A new scale for the measurement of interpersonal trust," *Journal of Personality*, 35: 651–665.

Scott, S.G. and Bruce, R.A. (1998) "Following the leader in R&D: the joint effect of subordinate problem-solving style and leader–member relations on innovative behavior," *IEEE Transactions on Engineering Management*, 45(1): 3–10.

Sears, G. and Holmvall, C. (2010) "The joint influence of supervisor and subordinate emotional intelligence on leader–member exchange," *Journal of Business and Psychology*, 25(4): 593–605.

Sluss, D., Klimchak, M., and Holmes, J. (2008) "Perceived organizational support as a mediator between relational exchange and organizational identification," *Journal of Vocational Behavior*, 73(3): 457–464.

Sparrowe, R.T. and Liden, R.C. (1997) "Process and structure in leader–member exchange," *Academy of Management Review*, 22(2): 522–552.

Stanley, A. (2006) *The Next Generation Leader: 5 Essentials for Those Who Will Shape the Future*, Multnomah, Sisters, OR.

Stringer, L. (2006) "The link between the quality of the supervisor-employee relationship and the level of the employee's job satisfaction," *Public Organization Review*, 6(2): 125–142.

Whittington, J.L., Goodwin, V.L., and Murray, B. (2004) "Transformational leadership, goal difficulty, and task design: independent and interactive effects on employee outcomes," *Leadership Quarterly*, 15(5): 593–606.

Yammarino, F. and Bass, B. (1990a) "Transformational leadership and multiple levels of analysis," *Human Relations*, 43: 975–995.

Yammarino, F.J. and Bass, B. (1990b) "Long-term forecasting of transformational leadership and its effects among naval officers: some preliminary findings," in K.E. Clark and M.B. Clark (eds) *Measures of Leadership*, Leadership Library of America, West Orange, NJ, pp. 151–171.

Yammarino, F. and Dubinsky, A. (1994) "Transformational leadership theory: using levels of analysis to determine boundary conditions," *Personnel Psychology*, 47: 787–811.

Yukl, G. (2009) *Leadership in Organizations*, 7th edn, Prentice-Hall, Englewood Cliffs, NJ.

Yrle, A.C., Hartman, S.J., and Galle, Jr, W.P. (2003) "Examining communication style and leader–member exchange: considerations and concerns for managers," *International Journal of Management*, 20(1): 92–100.

Part III
Assessment

7 Employees and the micro-level of sustainability performance

The benefits of sustainability

We have underscored in previous chapters the importance of leadership in helping to build sustainable organizations. Leadership is certainly critical in helping to create the organizational culture and social context that is necessary if firms are to derive the benefits associated with improving their environmental and social performance. By leveraging leadership, firms stand to benefit from improving their overall effectiveness, reduce organizational costs, and improve their innovative capabilities. Each of these is an important organizational outcome. However, these outcomes may overshadow an additional host of internal benefits that firms can derive from sustainability.

During the past decade or so, there has been increasing pressure on firms to engage in more socially and environmentally conscious activities. Much of this pressure has come from external stakeholder groups, including investors and consumers (Donaldson and Preston, 1995). As a result, a significant body of research has developed to understand how organizations have responded to these external pressures, and the potential range of benefits they can see by becoming more sustainable. For instance, research has investigated the monetary rewards associated with social responsiveness (Griffin and Mahon, 1997). Others have studied relationships between corporate social responsibility (CSR) and corporate financial performance (Alexander and Buchholz, 1978).

However, firms have also felt pressure from internal organizational constituents, such as managers and employees, to become more socially and environmentally responsive (Donaldson and Preston, 1995). In response, firms are beginning to see a number of surprising benefits when they meld sustainability into their overall strategic plans and include employees in the formulation and implementation of those initiatives. In this chapter, we review the array of benefits associated with making employees integral components and building blocks of company sustainability initiatives, and why employees play such a critical role in helping to improve the social and environmental performance of organizations. To underscore the importance of employees, we dig into the literature to provide leaders with an understanding of the reasons why social and environmental initiatives are salient factors to employees, and how sustainability can play a key role in improving employee engagement. We then highlight a number of examples, drawn from both large and smaller organizations, that have used sustainability in creative ways to improve employee engagement. We conclude with a list of challenges firms are likely to encounter on the road to becoming more sustainable organizations, and some key steps managers should consider when they attempt to engage employees through sustainability.

There are a number of ways in which sustainability can benefit organizations. Perhaps most importantly, sustainability can improve a firm's performance. The results of a recent National Environmental Education Foundation (NEEF) study found that sustainability can increase organizational profits by as much as 38% (NEEF, 2009). While firms can take a number of steps to help ensure positive financial outcomes are realized, a growing body of research has been devoted to the importance of employees in social and environmental responsiveness. We highlight the role of employees below.

Attracts talent

Research has shown that sustainability helps attract employees and managers to work in firms, especially those who seek out those organizations committed to environmental responsiveness (Moskowitz, 1972; Turban and Greening, 1996; Greening and Turban, 2000; Peterson, 2004). In addition to being attractive to job applicants today, sustainability helps firms pick and choose their future employees from a larger pool of job candidates (Turban and Greening, 1996; Greening and Turban, 2000).

Opinion surveys document support for these findings from academic research and indicate that the appeal of sustainable organizations is quite widespread in today's labor market. For example, a 2008 survey commissioned by *National Geographic* magazine found that over 80% of US workers believed it was important to work for an organization that makes the environment a high priority (National Geographic, 2008). Another survey, conducted by Harris Interactive National Quorum, revealed that over 60% of workers felt that a firm's impact on the environment was vital when evaluating a new workplace (Harris Interactive National Quorum, 2011). Interestingly, this is roughly the same percentage of workers in that survey who felt a potential employer's profit margin was vital. Further, over 70% of respondents believed that sustainability was an important or very important criterion in their job-search process. The study also revealed gender disparities with respect to how employees evaluate potential employers. For example, roughly half of male workers believed that a company's impact on the environment was important in their job-search process. In contrast, 78% of female employees felt that a company's impact on the environment was particularly important. The results of these studies underscore the importance today's labor market places on sustainability during the job-search process.

Improves loyalty and commitment

High-level executives know that the long-term business plans of organizations have a considerable chance of success when employees are loyal and committed to the firm. Unfortunately, what often happens in practice is that employees find themselves in organizations that expect loyalty, yet the firm does not develop the kind of environment and atmosphere that fosters loyalty among its employees. Sustainability not only helps organizations attract key employees, it also has the potential to improve employee/employer bonds in positive and meaningful ways. The ways in which sustainability can improve employee loyalty and commitment are discussed below.

Considerable research has demonstrated the effects of environmental and social responsiveness on employees' organizational commitment. Studies by Moskowitz (1972); Turban and Greening (1996); Albinger and Freeman (2000); Backhuas *et al.* (2002); and Peterson (2004) have all shown that a firm's social and environmental contributions not only attract potential employees, but can also improve the commitment levels of current rank-and-file employees. Greening and Turban (2000) also found that job applicants' and current employees' perceptions of a firm's social and environmental posture determines the organization's attractiveness to them. More recently, Brammer *et al.* (2007) noted that the social and environmental initiatives of firms increased employees' organizational commitment. That study found that the contribution of a firm's social and environmental initiatives to the organizational commitment levels is as important to employees as job satisfaction.

Employee loyalty helps contribute to greater efficiency, better business results, firm growth, reduced employee turnover, etc. (Meyer and Allen 1997; Antoncic and Hisrich 2004). Loyal employees can also help their firm develop and build the image it presents to outside stakeholders, including customers (Meyer and Allen 1997: 3). Firms should therefore make a concerted effort to evaluate how they can minimize coercion with respect to their employees and focus their efforts upon creating an environment where employees become committed and are willing to accept responsibilities.

There are numerous case examples of firms that have leveraged the environmental interests of their employees. The most positive outcomes have been among firms that leverage the talents of environmentally focused employees to form "green teams."

Develops employee skills and improves internal culture

Incorporating employees into their overall sustainability goals has a number of additional benefits for firms. Along with improving the loyalty and commitment of existing employees, and helping the firm attract top talent, firms that adopt a more environmentally and socially responsible posture can also experience positive environmental changes within the organization. Firms that strive to become more sustainable stand to gain by creating a healthy internal culture – a culture with an agreed-upon common goal that seeks to improve the performance of the organization while minimizing its negative impact on society and the natural environment.

Sustainability provides what many in today's workforce are searching for. Sustainability enables employees to see and potentially experience a deeper meaning and sense of purpose in their firm. Through the increased opportunities to have a positive impact on the world and to work in an organization with a good reputation in the community, sustainability offers employees the opportunities to develop knowledge and skills and advance their careers. For instance, researchers have found that firms can enhance employees' willingness and commitment by creating a sense of belonging through the volunteer efforts of employees in social and environmental initiatives (Lantos, 2002). When employees volunteer, they stand to improve their skills and training, and volunteering encourages teamwork among employees.

Improves the brand

Sustainability can help firms build brand awareness and, potentially, increased market share. Leveraging employees can help organizations achieve these outcomes by identifying new opportunities to improve their manufacturing and distribution practices to help achieve sustainability goals. In his recent book *Green Recovery*, Andrew Winston notes that "The engaged workforce will find more opportunities to get lean and identify more opportunities to innovate and create products and services that lower customers' environmental impacts. All of this work will improve the top and bottom lines" (Winston, 2009). In addition to helping build the brand internally, sustainability also can lead to employees feeling more secure and proud of their company, and being more willing to share the sustainability story of their firm with their friends. Each of these results helps build the brands of those firms that adopt sustainability initiatives.

Key principle

Sustainability provides a number of internal benefits to firms, including the attraction, retention, and skill development of the workforce.

Employees: keys to sustainability

The preceding section demonstrates the importance of employees to organizational efforts to improve sustainability. Employees are likely to join and remain with firms they feel are making a conscious effort to reduce their environmental footprint and to improve society. However, as important as sustainability is to bringing talented individuals into the organization, this relationship can also work in the opposite direction. Firms should understand that their employee base is critical to helping them achieve high levels of social and environmental performance. Hence the task facing high-level executives is to seek and determine the most effective ways to engage their employees, who are a key resource in helping the firm become more sustainable.

Studies have shown that high-engagement organizations tend to have higher operating income, revenue, earnings per share growth rate, retail sales, profit margins, customer retention, safety, and employee retention. However, major research firms such as TowersWatson, Gallup, and Blessing White have found that, across a wide variety of organizations, only 20–30% of employees are actively engaged. Given the importance of employee engagement to a host of positive organizational outcomes, these surveys help underscore the importance of employee engagement as one of the most important priorities facing managers today.

While there are a range of ways to motivate and inspire employees, sustainability offers an exciting and viable means through which managers can engage their employees. Sustainability offers a new and unique way to latch onto the passion and initiative of employees in their search for meaning in their work. Firms are likely to find that employee engagement and sustainability is quite interconnected, and can help launch dramatically improved business performance and potentially begin innovative breakthroughs. There are a number of reasons for this. Most importantly, sustainability offers what many in today's workforce are searching for – a deeper meaning to their work and a greater sense of purpose. By leveraging the resources of a firm to contribute and make a positive impact on the world, sustainability can serve as the means through which employees can meet their own personal desires to improve the environment and improve society. Employees are likely to respond positively to working in an organization that is striving to have a positive environmental and social impact and to improve its reputation in the community in substantive ways.

In previous chapters, we noted the growing evidence that sustainability is a business imperative. Most importantly, it can yield cost savings, enhances a firm's reputation, and offers the means through which new product and service innovations can occur, even in the most mature industries. However, most firms are a long way from reaping the benefits of sustainability. Much of the reason for this stems from their inability to recognize the business case for sustainability. Perhaps the primary reason why firms are slow to begin seriously evaluating how they can reduce waste and water/energy use stems from a lack of awareness about the ways in which employees can serve as a catalyst in helping the firm start down the path of sustainability, which is good for both the environment and the firm.

Poor communication is one factor that hampers employees' engagement in the sustainability activities of firms. Today, too many firms focus their communication about their sustainability efforts on external stakeholders (customers), while overlooking the importance of delivering information about their energy- and water-reduction strategies to their internal stakeholders (managers and rank-and-file employees). For example, a recent study found that social and environmental programs are not adequately communicated and implemented, and that 86% of current employees were not engaged by their company's sustainability program (Brighter Planet, 2009). The study also found that over 60% of respondents wanted to learn more about their employers' and co-workers' sustainability efforts; and 67% would like their employers to change their stance on sustainability. In a separate study conducted by IBM, only 46% of companies actively engaged front-line managers, and only one-third of companies engaged employees on corporate social and environmental objectives and initiatives (IBM, 2009).

Engaging employees through sustainability is just as important to firms that are leading the way in designing and implementing social and environmental initiatives as it is to those firms that are new to the "sustainability frontier." The good news is that the most innovative leaders are looking beyond green teams and other practices that confine and relegate sustainable thinking and idea generation to just a select few. Instead, firms that integrate sustainability across the value chain and embed sustainability thinking into all their organizational process are leading the field. Getting employees engaged in this process requires purposeful and focused communication and cooperation across divisions and business units. It means inspiring employees with a vision and a sense of purpose. Successful leaders are building systems and platforms that facilitate sustainability thinking such that it becomes an integral part of everyone's job, not an add-on activity that employees engage in during non-peak periods. Getting employees to engage in the firm's sustainability initiatives, whether the efforts are advanced or in their infancy, requires that leaders leverage all forms of communication. For instance, firms should incorporate social media and IT-enabled solutions to bring all of their employees, both foreign and domestic, to the table to help identify where water and energy is wasted within the firm, and to generate ideas on how both can be more closely monitored and their impacts reduced over time. Engaging employees in social and environmental solutions also goes beyond routine company meetings. Engaging employees in sustainability includes helping to design new approaches and models for volunteering and community investment, health and wellness, and community economic development.

Key principle

Employee involvement is essential if firms are to attain performance outcomes from their sustainability efforts.

Sustainability and employee engagement

There are a number of underlying reasons why sustainability will improve employees' engagement. Research has shown that employees' attitudes and behaviors are heavily influenced by how fair they consider their organization's actions to be (Cropanzano *et al.*, 2001). Moreover, employees have been found to react emotionally, both attitudinally and behaviorally, when they learn about an injustice directed toward somebody else, even when they do not identify with the victims (Folger *et al.*, 2005). CSR research has shown that job applicants' and employees' perceptions of a firm's CSR affects how attractive these employees consider the firm to be (Greening and Turban, 2000).

Social identity theory suggests that individuals will try to associate themselves with firms they consider to have attractive attributes and positive reputations (Brammer *et al.*, 2007). Individuals develop an attachment to their firm, which they can rely on to fulfill their needs of gratifying the self, enabling the self, and enriching the self. In other words, according to social identity theory, employees capitalize on the firm's organizational identity in order to establish, maintain, and enhance their own identity (Dutton *et al.*, 1994). Research has also shown that high levels of employee–employer identification result in an increase in affective commitment (Brammer *et al.*, 2007).

Based on this, employees may favor an organization that is concerned about its effects on society and the environment. Social and environmental initiatives have the potential to help employees enrich themselves, as working for a responsible company may help employees to enhance their own self-image and satisfy self-expressive needs, as well as the need for a meaningful existence (Rupp *et al.*, 2006). In addition, employees may also look positively on firms that demonstrate a willingness to establish policies and mechanisms, and even curtail certain practices, to improve society and the environment with which the firm comes into contact. These actions that the firm directs toward society and the environment may signal to employees that their firm cares for them personally, and they will have their needs met. Through sustainability initiatives, firms stand to benefit by building strong connections with their employees.

Challenges in engaging employees

A key factor facing organizations today is that most employees are not actively engaged. Sustainability offers leaders a viable opportunity to engage today's workforce. However, if firms hope to leverage the power of employees to help fulfill their sustainability objectives, leaders should give considerable thought to how this can be accomplished.

Measurement

In order to identify success, it is important to develop performance measures that will stretch your business units, yet still be attainable. Software tools are available that enable managers to track sustainability results at the employee team level. For instance, AngelPoints (www.angelpoints.com), a provider of software tools for employee engagement, has partnered with the sustainability strategy firm Saatchi & Saatchi S (www.saatchis.com) to create a web-based platform that helps firms engage their employees in sustainability and follow their progress in reducing their personal carbon footprint, improving their health and wellness, and reducing waste.

Communication

Not all your employees will be engaged in your firm's sustainability efforts. Therefore it is critical that managers, especially front-line and mid-level managers, communicate to employees, in meaningful ways, the salience of the firm's sustainability efforts. Moreover, firms should not rely on their published sustainability reports as their primary means of communicating their sustainability initiatives to their employees. Firms should update employees directly and frequently about their sustainability initiatives, and about how they tie back into the firm's overall strategic plan.

Sustainability as in-role behaviors

In order for a firm's sustainability efforts to take hold at all levels of the organization, it is imperative that managers expect and assist employees to formulate and implement sustainability activities in their day-to-day jobs – the employees' in-role job behaviors. Optimally, employers should strive to help their workforce embrace the sustainability initiatives of the firm to such an extent that employees do not consider sustainability efforts to be outside the scope of their normal job role. In doing so, employees will clearly demonstrate in-role sustainability behaviors, such as building water, energy, and resource savings into their day-to-day activities.

Most importantly, it is essential that managers meet regularly with their direct employees, both formally and informally, and solicit input on the ways in which each employee is implementing sustainably in their particular job. In addition, sponsoring educational events from environmental professionals and government agencies can help underscore the opportunities employees have to reduce their own environmental footprint while averting unnecessary energy or materials use. Building an internal dialogue among employees and managers about sustainability best practices in each job across the firm gives employees a sense that the activities of the entire workforce regarding sustainability are worthwhile and are supported by all areas and levels of the firm.

To underscore to employees the importance the firm is placing on sustainability, regular manager-to-employee dialogue about sustainability should also be reinforced by a firm's performance-review process. Embedding sustainability into performance evaluations signals to employees that they are recognized for the day-to-day activities they implement that help the firm to improve its social and environmental performance. We highlight below several examples of firms that have embedded sustainability in the foundation of their organization and within the job roles of employees at all levels.

Key principle

Engaged employees incorporate sustainability into their day-to-day job, demonstrating clear in-role sustainability behaviors.

Sustainability as extra-role behaviors

If a firm is to have a workforce that is engaged in its sustainability efforts, it should seek to establish a corporate culture where employees seek out ways to improve the firm's environmental and social performance beyond their day-to-day role – the employees' extra-role behaviors. Research has shown that successful sustainable operations depend on the voluntary organizational citizenship behaviors (OCBs) of employees (Boiral, 2009).

While there are potentially many ways to accomplish positive sustainability-related extra-role employee behaviors, firms should actively build a climate that encourages employees to engage in helping behaviors centered on improving the social and environmental performance of the organization. These extra-role behaviors, or activities outside employees' codified job duties, provide firms with the energy and momentum needed to build commitment within teams and organizational units around the organization's sustainability efforts. There are several ways in which a firm can encourage extra-role sustainability behaviors among its workforce. First, management can ask employees to contribute to shaping the firm's sustainability efforts by providing opportunities for them to offer input regarding the design of the firm's sustainability strategy. Second, management can solicit employees' input about suggested sustainability actions for departments outside their own. Third, employees can be offered opportunities to become ambassadors for the firm's sustainability efforts in sustainability training and communication sessions with other employees. Stonyfield Farm, Burt's Bees, and IKEA (described in the following section) are just some examples of firms that encourage employees to engage in sustainability-related activities outside their normal job duties.

Employees engaged in their firm's sustainability efforts can also lead to an aspect of extra-role behaviors that we term community citizenship behaviors (CCBs). Examples include Hershey's (www.hersheys.com), which offers its employees support for volunteering in their local community. At Solo Cup Company (www.solocup.com), over 300 employees participated in more than forty-five recycling, education, and beautification events in the USA and Canada. This was a 110% increase over 2009, when Solo first launched its sustainability action network. Through the network, volunteers identify local or company-wide projects and lead the way in executing them. Alcoa (www.alcoa.com) employees volunteer in their local communities through the Alcoa Green Works initiative to support environmental projects and celebrate eco-holidays such as Earth Day, World Environment Day, and Arbor Day. At Caterpillar (www.caterpillar.com), more than 100 employees and their families volunteered to clean up trash along the Illinois River, in a joint effort with Living Lands & Waters, a local nonprofit organization dedicated to the protection and conservation of America's rivers. Campbell Soup (www.campbellsoup.com) expanded national Make a Difference Day to Make a Difference Week four years ago, after a record number of employees expressed interest in volunteering. According to a General Mills survey, 82% of the company's US employees volunteer either through company programs or independently, and nearly 60% of employees spend up to five hours a month serving in their community (General Mills, 2011). Each of these examples demonstrates the willingness of engaged employees to go beyond their in-role job activities, and even to go beyond their extra-role OCBs, by demonstrating community citizenship behaviors through volunteering and seeking out company-sponsored socially and environmentally beneficial activities that help improve society at large.

Key principle

Engaged employees incorporate sustainability beyond their day-to-day jobs, demonstrating clear extra-role (OCB and CCB) sustainability behaviors.

Cases of employee engagement and sustainability

Recent reports by the HR consultancy Towers Watson found that companies with engaged employees demonstrate significantly higher operating income, employee productivity, and five-year shareholder return. Not surprisingly, they also show dramatically lower costs associated with employee turnover and retraining (Towers Watson, 2011). Today, companies are becoming increasingly aware of the tangible ways in which their sustainability efforts around the world can pay dividends both for the community and for organizations. Leadership is critical in order to establish social and environmental initiatives. However, employees' passion is just as vital to help formulate and implement these initiatives.

One way in which firms can engage employees through sustainability is through "green teams." In a recent study entitled "How to build successful green teams," researchers examined fourteen leading firms and identified a number of common themes among successful teams (Sabre Holdings, 2010), including the following.

- Green teams that did not utilize methods for effective collaboration were less successful than teams that did use these methods.
- Even if a green team is not successful in innovating new products or services, the processes in which the team engages may still have a positive impact on organizational culture by connecting, engaging, and inspiring employees to have a shared commitment towards the environment.
- Green teams often have more success when they include a representative from each business unit. By leveraging multiple and diverse perspectives, the team can make better plans for the future.
- Perhaps the most important key to a successful green team is executive support.
- Several of the companies surveyed used either formal or informal processes to capture ideas from employees that drive product and service innovations,

increase operational efficiencies in the workplace, and keep the business on the cutting edge of its industry.

- Green teams are most often organized at a grassroots level. People who are passionate about sustainability and see opportunities in their workplace to make a positive difference organize groups to make changes based on their observations.

Another recent study, "Green teams and value: engaging employees in meeting sustainability goals" (AltaTerra Research, 2010), also addresses how green teams can contribute effectively to a firm's organizational environmental goals; the value or return on investment they can help to deliver; and how they are best organized to do so. Among the companies highlighted in the report are Lockheed Martin, Palm (recently acquired by HP), Genetec, and eBay. The report outlines several key factors for program success, including bidirectional communication, effective decision-making processes, organizational connectivity, maintaining enthusiasm, and measuring success. The results suggest that leading firms are organizing around these key capabilities to create successful support structures for green team programs.

Company examples

Burt's Bees

Burt's Bees has a mandatory sustainability training program that all employees must take for up to thirty hours per year. The company was recently a top award winner in the UK daily *The Guardian*'s "Sustainable Business Awards: Engaging Employees." Burt's Bees' thirty-hour requirement allows its employees to partake in training programs devoted to environmental stewardship, social outreach, natural wellness, and leadership.

IBM

IBM invites its employees to help determine the company's overall sustainability strategy. The firm's "Big Green Innovations" program includes environmental initiatives focused on advancing water management, alternative energy, and carbon management. The program was formulated as a result of the 2006 IBM innovation jam. Among the 30,000–40,000 new ideas offered by employees, the company decided strategically to adopt the Big Green Innovations program.

IKEA

IKEA stores have both an environmental coordinator and an action plan for recycling, waste sorting, energy saving, transport of goods, and education. In addition, the company focuses on education and transparency with employees about the firm's sustainability initiatives. For instance, IKEA offers online and classroom training focusing on material relating to the company's sustainability direction, why the employees are receiving sustainability training, and information on global environmental issues such as climate change. The firm also shares with employees the environmental and social areas it aims to focus on, the environmental coordinator and committee structure, and recycling and environmental management, along with practical examples. The firm is also transparent with employees regarding its procurement standards, its approach to forestry, partnerships, and social projects on a global level.

Best Buy

Best Buy relies on social media to listen to and engage its employees about sustainability. Mary Capozzi, Best Buy's senior director for corporate responsibility, states that investing in employee education and training is vital. "Our employees are crucial in our ability to deliver sustainable solutions to our customers. They link what we do as a business with the need of our customers." In addition, Best Buy focuses on transparency with its employees regarding social and environmental issues. The firm accomplishes this by creating a sustainability and corporate responsibility scorecard. The scorecard is designed to give Best Buy employees an opportunities to see the firm's social and environmental metrics and to help them understand how their efforts have a direct effect on Best Buy's ability to meet their sustainability objectives. Best Buy's scorecard tracks the firm's social and environmental performance in four key areas: community giving and volunteer efforts, recycling, store energy performance, and sales of Energy Star-rated products. It is important also to note how often Best Buy communicates with its employees about sustainability. Rather than relying solely on an annual sustainability report, Best Buy uses an internal magazine, *The Link*, that goes to each of Best Buy's stores six times a year, to include stories about how store teams are contributing to helping Best Buy improve its sustainability.

BT (formerly British Telecom)

This company's "Great Switch Off" campaign aims to help raise awareness among employees about energy use and climate change by encouraging them to turn off computers and lights around the company. The firm uses posters in the cafeteria, coffee rooms, and break rooms, and also advertises via email. Volunteer floor captains audit electronics and office lighting, then follow up with a second audit in a month. Floor captains leave stickers if computers are left on standby or lights are left on. This process helped reduce the number of electronic items left on standby by 20% over the month of the first Great Switch Off. The floors that saw the greatest energy use reduction received a gift.

Stonyfield Farm

Stonyfield uses the hiring process to focus employee engagement and education on sustainability issues. Its "Mission Action Plan" engages all the firm's employees in its sustainability mission by linking long-term environmental impact goals to job performance measures.

Hewlett-Packard

HP connects with employees by utilizing mixed-media communications, events, and programs. The firm provides sustainability education to employees through brown-bag information seminars about installing solar roofs at their own homes, and incentives for employees using solar energy. The firm also supports grassroots employee sustainability networking.

Cisco Systems

Cisco has historically maintained a culture of entrepreneurship, innovation, and collaboration, and has extended this to the environmental arena. The firm's commitment to sustainability began when a small group of employees formed a cross-functional team to address environmental concerns and opportunities. In 2006, Cisco established an EcoBoard made up of fourteen leaders from across the company.

US Postal Service

The US Postal Service has also benefited from engaging employees in sustainability. Its "Lean Green Teams" focused on reducing energy, water, solid waste to landfills, and petroleum fuel use, helping to save the agency more than $5 million in 2010. In addition, these employee teams helped the Postal Service recycle more than 222,000 tons of material – an increase of nearly 8000 tons over the previous year – which generated $13 million in revenue and saved an additional $9.1 million in landfill fees.

Alcoa

Alcoa demonstrates the importance of executive support, with 20% of executive compensation based on progress toward specific sustainability goals. The firm distributes an annual sustainability report to reinforce accountability, and has created employee volunteer initiatives including "Green Works," "Alcoa Earthwatch Fellowships," and "Make an Impact" with the Pew Center on Global Climate Change.

Unilever

Unilever began its "Vitality Programme" in 2005 as a way to encourage employees to lead healthier lifestyles. While Unilever's program offers many benefits to employees, its goals are to promote the wellbeing of employees in terms of fitness of body, heart, mind, and spirit.

Intel

Intel uses its "Involved Program" to enable employees to volunteer thousands of hours in the communities where they work.

Accenture

Accenture allows employees to work on non-profit consulting projects in developing countries through its "Accenture Development Partnership" program.

Key principle

Sustainability provides employers many opportunities to build stronger bonds with employees.

Keys to helping engage employees with sustainability

There are a number of ways in which firms can increase employees' engagement through sustainability.

Education

Education is increasingly being relied upon by firms that want not only to increase employees' awareness of environmental and sustainability issues, but also to encourage employees to generate ideas regarding how the organization can become more sustainable. Through education and awareness, firms hope that employees will adopt more sustainable choices and behaviors at work. For example, Johnson & Johnson has created environmental literacy programs aiming to change employees' viewpoints or feelings toward the environment. According to the company's website, over 90% of its facilities have implemented a literacy campaign centered around discussing "successes and progress and going after their hearts a bit." Johnson & Johnson claims the program has been a "great success among employees."

Incentivize sustainable behaviors

Incentives are another means through which firms can engage employees through sustainability. In 2008, Stonyfield Farm recognized that it needed to take significant steps to decrease energy consumption in its facilities. One of its strategies was to reward employees for engaging in sustainable practices. By tying funds saved from decreased energy and water usage directly to paycheck bonuses, Stonyfield was able to decrease usage by more than 22% that year (NEEF, 2009).

Linking sustainability to corporate goals

This is a key component in helping employees recognize the importance of sustainability to the firm, and in building a bond with employees. Leaders who purposefully tie reductions in energy, waste, and water usage into the overall framework of corporate strategies signal to employees that the firm is serious about sustainability and considers it not as a tangential factor, but rather as a key to the firm's ongoing success.

Communication

Communication is vital if firms are to expect employees to recognize the importance of sustainability to the organization and how it affects their current role, and willingly to step forward with innovative solutions to reduce waste, water, and energy usage. As our case examples above demonstrate, there are a number of ways in which firms can do this. While all forms of electronic and social media should be leveraged, firms should not overlook strategically located posters and fliers throughout the organization. Firms should also incorporate the importance of sustainability into the ongoing dialogues that occur in the firm, such as production meetings and weekly operational staff meetings.

Recognize achievement and participation

Recognition and rewards help underscore the importance of sustainability to employees, and build an understanding among employees that making sustainable decisions and acting in sustainable ways leads to rewards within the organization. In addition to simply rewarding employees who engage in sustainable activities, whether individually or collectively, firms should consider carefully the processes through which these rewards are decided upon and how they are distributed. If firms are sincere in their efforts to change the mindsets of employees and build the type of bonds that encourage employees to engage fully in their organization, they should incorporate high-level management in the sustainability recognition and achievement process.

Key principle

While sustainability presents employers with a number of opportunities to engage employees, they should consider carefully how to reward sustainable behaviors.

Summary

- Firms derive many internal benefits through sustainability.
- Sustainability helps attract talent, improves loyalty and commitment, develops employee skills, and improves internal culture.
- Engaging employees in and through sustainability efforts is just as important to firms that are leading the way in designing and implementing social and environmental initiatives as it is to those firms that are new to the sustainability frontier.
- There are many challenges in engaging employees with sustainability, including communication and developing awareness.
- Many firms are recognizing the business case for sustainability, and leveraging their employee base to help them identify ways to reduce their environmental footprint and improve their social performance.
- Firms that use sustainability to improve employees' engagement can experience benefits through improved employee in-role and extra-role (OCB and CCB) sustainability behaviors, both of which can enable firms to reduce waste, energy, and water usage.

Discussion questions

- What are some of the internal benefits that firms derive through sustainability?
- Describe the challenges associated with engaging employees.
- Explain how social identity theory can help inform managers of the ways in which sustainability can resonate with employees.
- What are the mechanisms firms can use to engage employees through sustainability?
- How can employees' engagement be measured?
- Have your colleagues demonstrated sustainability in their current job roles?
- Are your employees aware of the volunteer programs that your firm sponsors? How can your firm coordinate with employees to identify and select the most important environmental and socially related needs in their communities?

References

Albinger, H.S. and Freeman, S.J. (2000) "Corporate social performance and attractiveness as an employer to different job seeking populations," *Journal of Business Ethics*, 35: 243–253.

Alexander, G.J. and Buchholz, R.A. (1978) "Corporate social responsibility and stock market performance," *Academy of Management Journal*, 21: 479–486.

AltaTerra Research (2010) "Green teams and value: engaging employees in meeting sustainability goals," www.altaterra.net/members/blog_view.asp?id=272897&post=110068

Antoncic, B. and Hisrich, R.D. (2004) "Corporate entrepreneurship contingencies and organizational wealth creation," *Journal of Management Development*, 23(6): 518–550.

Backhuas, K.B., Stone, B.A., and Heiner, K. (2002) "Exploring the relationship between corporate social responsibility and employer attractiveness," *Business and Society*, 41: 292–318.

Boiral, O. (2009) "Greening the corporation through organizational citizenship behaviors," *Journal of Business Ethics*, 87: 221–236.

Brammer, S., Millington, A., and Rayton, B. (2007) "The contribution of corporate social responsibility to organizational commitment," *International Journal of Human Resource Management*, 18: 1701–1719.

Brighter Planet (2009) *Employee Engagement Survey: An Analysis of the Extent and Nature of Employee Sustainability Programs*, Brighter Planet, Middlebury, VT and San Francisco, CA. http://attachments.brighterplanet.com/press_items/local_copies/55/original/employee_engagement_2009.pdf?1265816076

Cropanzano, R., Rupp, D.E., Mohler, C.J., and Schminke, M. (2001) 'Three roads to organizational justice," *Research in Personnel and Human Resources Management*, 20: 1–113.

Donaldson, T. and Preston, L.E. (1995), 'The stakeholder theory of the corporation: concepts, evidence and implications," *Academy of Management Review*, 20: 65–91.

Dutton, J., Dukerich, J., and Harquail, C. (1994) "Organizational images and member identification," *Administrative Science Quarterly*, 39: 239–263.

Folger, R., Cropanzano, R., and Goldman, B. (2005) "What is the relationship between justice and morality?," in J. Greenberg and J.A. Colquitt (eds) *Handbook of Organizational Justice*, Lawrence Erlbaum Associates, Mahwah, NJ, pp. 215–245.

General Mills (2011) "Volunteerism," www.generalmills.com/en/Responsibility/Community_Engagement/Impact_on_society/Volunteerism.aspx

Greening, D.W. and Turban, D.B. (2000) "Corporate social performance as a competitive advantage in attracting a quality workforce," *Business & Society*, 39: 254–280.

Griffin, J.J. and Mahon, J.F. (1997) "The corporate social performance and corporate financial performance debate," *Business and Society*, 36: 5–31.

Harris Interactive National Quorum (2011) "U.S. workers declare employers' dedication to environment as important as profit when evaluating a new workplace," Interface, Inc., Atlanta, GA, www.harrisinteractive.com/vault/Interface-Employee-Environment-04-04-11.pdf

IBM (2009) "Corporate social responsibility: leveraging insight and information to act," IBM, Armonk, NY, www-935.ibm.com/services/us/gbs/bus/html/csr-study-2009.html

Lantos, G.P. (2002) "The ethicality of altruistic corporate social responsibility," *Journal of Consumer Marketing*, 19(2): 205–230.

Meyer, J.P. and Allen, N.J. (1997) *Commitment in the Workplace: Theory, Research and Application*, Sage, Thousand Oaks.

Moskowitz, M. (1972) "Choosing social responsible stocks," *Business & Society*, 1: 71–75.

National Geographic (2008) "U.S. workers favor green companies," http://press.nationalgeographic.com/pressroom/index.jsp?pageID=pressReleases_detail&siteID=1&cid=1203526072027

NEEF (2009) *The Engaged Organization: Corporate Employee Environmental Education Survey and Case Study Findings*. National Environmental Education Foundation, Washington, DC.

Peterson, D.K. (2004) "The relationship between perceptions of corporate citizenship and organizational commitment," *Business & Society*, 43: 296–319.

Rupp, D.E., Ganapathi, J., Aguilera, R.V., and Williams, C.A. (2006) "Employee reactions to corporate social responsibility: an organizational justice framework," *Journal of Organizational Behavior*, 27: 537–543.

Sabre Holdings (2010) "Generating sustainable value: moving beyond green teams to transformation collaboratives," Sabre Inc., Dominican University and Paladin Law Group LLP, www.greenbiz.com/sites/default/files/Generating%20Sustainable%20 Value%20Through%20Employee-led%20Teams%20Final.pdf

Towers Watson (2011) *Perspectives: The Power of Three: Taking Engagement to New Heights*, Towers Watson, New York, www.towerswatson.com/research/3848

Turban, D.B. and Greening, D.W. (1996) "Corporate social performance and organizational attractiveness to prospective employees," *Academy of Management Journal*, 40: 658–672.

Winston, A. (2009) *Green Recovery: Get Lean, Get Smart, and Emerge from the Downturn on Top*, Harvard Business Press, Cambridge, MA.

8 Organizations and the macro-level of sustainability performance

Organizational benefits of sustainability

Today, managers are facing pressures from a growing number of stakeholder groups to lead their firm to take on more sustainability initiatives and to take a more active role in improving their social and environmental performance records. Many managers are responding to these pressures by searching for ways to lead their firms to become more sustainable. Despite a great deal of effort, managers often struggle to identify ways in which they can properly evaluate the outcomes of their sustainability initiatives – understanding the benefits associated with sustainability is a challenging task facing managers today. Acquiring the information needed to justify investing in sustainability initiatives can be quite difficult when organizational leaders do not understand and appreciate the full range of benefits that sustainability can bring. Without a clear understanding of sustainability benefits, managers are apt to see sustainability as nothing more than activities that are outside their core business and will only add costs.

In this chapter, we present a variety of ways in which leaders can assess the environmental and social performance of their organizations. We show that these assessments can be made internally, using performance tools that add social and environmental dimensions alongside economic performance metrics. In the first section, we outline the many ways in which sustainability can help firms. We then discuss ways in which firms can comprehensively evaluate their environmental and social performance, through the triple bottom line, the balance scorecard, and returns on investment in sustainability-related initiatives. Each of these can provide leaders with the means to assess the strategic and operational outcomes of their sustainability initiatives.

We then discuss, in progressively more detail, the finer points of sustainability performance measurement. We outline what firms should measure, and how they should rely on the principles of external groups when deciding whether and how to report certain organizational features that affect the environment. In this section, we outline a number of sustainable metrics on which leaders can draw, including those from external sources. In the remaining sections, we discuss the importance of third parties to sustainability performance. Along with properly measuring sustainable performance, leaders should recognize that another means through which sustainability improves organizational performance is by enhancing the reputations of companies. Over the past ten years we have seen significant growth in the number of groups rating the sustainability efforts of firms. Much has changed in that period, including some rating agencies merging with others. For example, over a fourteen-month time span, RiskMetrics purchased Innovest (February 2009), then purchased KLD (November 2009), then itself was purchased by MSCI (March 2010). Today, many organizations are taking advantage of the reputational benefits derived from the certifications of these rating agencies.

In this chapter we demonstrate that there are, in fact, many opportunities for leaders to see performance gains by adopting the pursuit of sustainable initiatives. However, outcomes of "green" leadership should be assessed with tools designed for that purpose. We identify both internal and external assessment tools that enable managers to see outcomes of their green leadership efforts.

Today, both investors and a wide array of corporate stakeholders have begun to take a keen interest in the sustainability of businesses. Executives are becoming aware of the need to incorporate environmental, social, and governance issues into their core business practices. By increasing media attention and government interventions towards sustainability issues, alongside an increasing awareness among consumers, the benefits associated with sustainability are becoming increasingly clear. Box 8.1 provides an overview of the benefits associated with improving firms' sustainability.

Box 8.1 Business benefits of sustainability

- Building competitive advantage
- Reducing costs
- Attracting and retaining skilled employees
- Enhancing employee citizenship behaviors and engagement
- Improving investor relations
- Improving customer loyalty
- Reducing litigation exposures
- Fostering innovation
- Leading regulation
- Fostering differentiation
- Encouraging growth and expansion
- Building the corporate brand

Sustainability: leads to competitive advantages

Along with traditional market factors including brand image, price, and value, the environmental and social credentials of firms are helping to establish niches and new markets and toppling traditional competitive logic. For example, a study by Information Resources Inc. found that as many as 20% of US consumers are "sustainability-driven," while 50% of US consumers consider at least one sustainability factor when making brand and store selection (Hespenheide *et al.*, 2010).

Some firms have capitalized on this in their product development activities. For example, Clorox has taken several steps to help decrease its environmental impact (Clorox, 2010b). The company's "Green Works" product line helped begin a profitable partnership with the Sierra Club. With this partnership the company's sales doubled, and $1.1 million was donated to the Sierra Club (Clorox, 2010a). Clorox has also focused the efforts of one branch of R&D to discovering innovative used of natural ingredients and recycled materials. The firm's "Eco-Office" brand includes compostable plastic bags and a pine cleaner made using the by-products of the paper and pulp industry. In addition, Clorox's goals have had cascading effects on internal stakeholders by helping foster a culture that embrace sustainability. Employees now conduct "dumpster dives" to identify cost-saving waste-elimination opportunities, and reduced organizational waste by 50% over one year (Clorox, 2010b). Clorox's sustainable goals helped propel it to be the first major household products maker to list detailed descriptions of all of its products' ingredients on its corporate website and to be recognized by the US Environmental Protection Agency (EPA) as a Safer Detergent Stewardship Partner. Clorox was also named one of the top 100 green companies in *Newsweek*'s rankings of S&P 500 companies (Clorox, 2010b).

Sustainability: reduces costs

One of the most important benefits firms can reap when pursuing sustainability initiatives is in the area of cost savings accruing through efficiency gains. Moreover, cost savings from your firm's sustainability efforts could develop new revenue streams. Whether your firms focuses its efforts on energy, water, waste, or raw materials, the metrics used to identify and reward savings can help identify new opportunities for savings.

Several companies have recognized how the environmental and social impacts of their products can present new opportunities. Shell's conflict with the citizens of Nigeria and allegations of Nike's labor practices demonstrate that sustainable operations can provide firms the opportunity to continue their license to operate in less developed and emerging markets. Others, such as Hewlett-Packard (HP), focus on eliminating waste both within its own operations and throughout the life cycle of its products. The firm helped one of its clients save over $3 million on managed print services over two years through a combination of recycling print cartridges, default duplex settings, and energy efficiency (Hewlett-Packard, 2010). HP also helped a European grocery chain save power consumption by 10–15% (Hewlett-Packard, 2009a). HP's organizational approach appears to prioritize savings wherever they arise and at any scale. For instance, in-house reuse and recycling policies helped to save the company $1.75 million in 2008. On the other hand, the design of a data center in Wynyard, UK that uses winds off the North Sea for building cooling is expected to save $15 million annually by using 40% less power than comparable facilities (Hewlett-Packard, 2009b). Each of these examples from HP shows the benefits that can come when firms dig deep into the full range of their operations for potential energy savings or carbon-reduction possibilities.

Sustainability: sparks innovation

Compliance with environmental demands and social welfare expenditures was viewed by organizations as a cost that very often correlated negatively with profit. Yet engaging in strategic sustainability initiatives can enable firms to derive competitive advantages over rivals. Environmental regulations, and the threat of pending regulations, can provide firms the impetus and motivation to derive innovative ways to reduce energy costs and water use in their supply chains as well as introduce innovative environmentally friendly products or services.

Sustainability: leads regulation

Coming up with innovative ways to meet or beat compliance targets associated with environmental regulation not only can reduce costs; it has also helped steer environmental regulation in directions to firms. Indeed, researchers argue that firms can "manage" their competitors "by imposing a set of private regulations or by helping to shape the rules written by government officials" (Reinhardt, 1999). For example, John Deere's technological innovation of compression wave injection was essential in driving industry standards for handheld lawn and garden equipment and creating a competitive advantage for John Deere because it exceeded the EPA's proposed standard at a lower cost than alternatives (CIM, 2007). In addition, John Deere was able to license the technology to its competitors, helping to make the cleaner burning engine more widely available and ensuring a future revenue stream.

Sustainability: develops intangibles

Intangibles related to environmental and social responsibility can also influence how customers and other stakeholders view and interact with firms. Improvements surrounding sustainability are likely to correspond to the ways in which stakeholders view the firm. For example, when Paul O'Neill (formerly secretary of the US Treasury) was the CEO of Alcoa, the firm's safety record was roughly one-third the national average. Rather than tackling the safety record through traditional measures, O'Neil focused on elevating health, safety, and environmental issues to a strategic level. By emphasizing such intangible values, O'Neill explained, "You will go beyond what seemed possible before. Workplace safety is just one example – when you become effective at organizational problem solving, many of the targets that seemed out of reach, whether that means profits, growth, innovation or new markets, become easy to grasp" (Potier, 2002).

Sustainability: improves risk management

Investments in environmental measures can help firms to limit downside risk, and help them to overcome public relations disasters. Organizations that use inputs in their production processes more efficiently and to meet environmental standards (as well as those likely to be enacted in the future) can also help to curtail the obsolescence of their products and services and mitigate rising production costs. Along with this, designing parts that are easier to upgrade rather than replace and dispose of can help extend product life and minimize obsolescence, ultimately improving the firm's standing with stakeholders.

Sustainability: improves shareholder accountability

Along with rising consumer interest in sustainability-related products, investors are increasingly interested in firms that are more socially and environmentally focused in their operations. Institutional investors such as the Investor Network on Climate Risk and the Carbon Disclosure Project are calling on firms to make greater efforts to disclose and account for their climate-change impacts. Likewise, the competition to be included on the Dow Jones Sustainability and FTSE4Good Indexes is becoming increasingly fierce. The US Securities and Exchange Commission recently published guidance that climate change-related risks could trigger a number of reporting requirements under federal securities law (SEC, 2010).

In response to stakeholder pressure, Nike has moved from a compliance-based environmental and social strategy to develop a more clearly defined vision of corporate responsibility that leverages innovation to promote profitability (Nike, 2010). The firm has strived to include all of its products' sustainability impacts into product designs. In addition to these steps, Nike saved $8.2 million in 2009 by better utilizing container use to cut emissions and unnecessary shipments (ibid.).

As the examples above illustrate, there are a range of organizational benefits associated with leading firms to improve their economic, social, and environmental performance. In the following sections we dig more deeply into how sustainability can pay off for firms.

Key principle

Organizations that pursue sustainability realize a number of positive tangible and intangible performance outcomes.

Comprehensive sustainability performance assessment

Today, organizations are under significant pressure to measure and report their social and environmental performance. In response, leaders are realizing that implementing sustainable business practices should go beyond simply creating good publicity. When managers conceive and implement sustainable initiatives properly, they can enhance firm growth while promoting social and environmental stewardship. However, measuring organizational performance is difficult, especially when what has to be measured keeps changing. Sustainability has widened the scope of what organizational activities should be measured, and as a result many leaders today are working on finding the most appropriate ways to measure the success of their sustainable initiatives. Measuring performance, whether economic, social, or environmental, is a key tool that managers can use for the control and implementation of initiatives. To assess levels of achievement, reassess priorities, and assign resources, managers need to have goals, measures, and objectives. In this respect, sustainability performance should be no different. In the following sections we highlight the importance of social and environmental performance measurement.

Without ways to establish an internal benchmark for continual improvement, it is quite difficult to innovate, advance, and proactively respond to stakeholder expectations with your firm's sustainability initiatives. There are a number of advantages to monitoring and measuring a firm's environmental performance. Much of the benefit firms can derive from linking environmental and economic performance is contingent on their ability to link environmental management practices into the normal course of a company's operations. The ability to quantify environmental performance in a meaningful way is critical to the effectiveness of this integration. When firms step up the scrutiny of their sustainability initiatives, they are likely to uncover a considerable amount of waste, which will likely equate to valuable opportunities to reduce costs. As Box 8.2 illustrates, there is a range of organizational benefits that firms derive when they measure sustainability.

Box 8.2 How sustainability measurement helps firms

- Create effective priorities
- Benchmark for improvement
- Encourage innovation
- Reinforce accountability across business units
- Improve goal setting and strategic planning
- Strengthen communication across business units

Companies that utilize best practices in measuring and improving environmental performance often include the entire life cycle of their products from the point of raw material extraction all the way through the production, distribution, recycling, and disposal phases. Nike pursued a holistic approach by taking an in-depth look at its shoe-manufacturing processes. By doing so, the firm realized that it was using three shoes' worth of material to produce one pair of shoes, resulting in a total cost of $700 million a year (Fromartz, 2009). After discovering this, Nike reinvented its design and manufacturing processes to focus on reducing waste. These steps helped Nike reduce waste by up to 67%, energy use by 37%, and solvent use by 80%. The company plans to convert all Nike shoes and clothing to meet higher environmental standards by 2011 and 2015, respectively (Fromartz, 2009).

Procter & Gamble (P&G) began recognizing the importance of sustainability when it started to link sustainability principles into its strategic plans and organizational culture more than a decade ago. The company became an industry leader by creating a formal measurement and control system that translates lofty sustainability objectives into tangible goals and strategies, and that communicates well to rank-and-file employees. For example, P&G's development of a new laundry detergent demonstrates value that is derived from its sustainability measurement and reporting system. P&G used a life-cycle assessment to measure the total environmental impact of the product, including basic ingredients, packaging, and disposal, along with a net present value analysis to determine the product's profitability (Busco *et al.*, 2010). Finally, P&G performed a social assessment to determine the product's benefits to consumers. Initially launched in the UK as Ariel, the new laundry detergent was an initial success in that it provided customers with an effective detergent that also saves them money through reduced energy consumption (ibid.).

The triple bottom line

To show quantitative improvement in economic, social, and environmental aspects, firms today are increasingly adopting the "triple bottom line" approach. The triple bottom line is based on the notion that an organization should measure its performance not only in relation to its shareholders, but also in relation to the firm's broader stakeholder constituency. The triple bottom line adds environmental and social performance measures to the financial measures managers typically use to assess firm performance.

- **Environmental performance** generally refers to the volume of natural resources, such as energy, water, and land, that organizations use in constructing and delivering their goods or services. It also refers to indirect results of firms' operations, such as waste and air emissions.
- **Social performance** is related to the general impact of a firm (including subsidiaries and suppliers) on the communities where it works.

Unfortunately, many executives are likely to find the triple bottom line approach to be troubling because it conveys that the firm's responsibilities are considerably larger they may have assumed. However, managers should understand that sustainable performance is more than the production of goods and services that customers demand at a profit.

Triple bottom line: measurement

Measuring the triple bottom line provides leaders with the ability to make better decisions regarding the competing costs and benefits of sustainability-related measures. Defining the return on investment, which is normally associated with financial metrics, can be extended to environmental and social activities. For example, savings due to more environmentally friendly operations and improvements in the firm's image in the eyes of stakeholder groups are all ways of measuring economic, social, and environmental performance.

Unfortunately, measuring social and environmental performance is not a straightforward task. Shareholder value, market share, customer satisfaction, even employee well-being, are relatively easy to quantify, and measures developed by one organization are readily transferable to others. Social and environment performance, however, are unique features of every organization, and they are often quite difficult to quantify.

Triple bottom line: environmental performance measurement

Some organizations have tackled the challenge of measuring their triple bottom line environmental performance by adopting internationally recognized, industry-certified environmental management systems (EMS). Such systems help firms develop, implement, and communicate environmental policies, set objectives and targets for reducing environmental impacts, and monitor performance against these targets. The leading EMS, ISO 14001, was introduced in 1996, and by 2005 over 36,000 ISO certificates had been awarded to organizations operating in 112 countries. ISO 14001 certifies that an organization has a certain type of EMS in place, and provides a general signal to organizational stakeholders that the firm will be actively managing its environmental impacts. The adoption of ISO certification appears to be accelerating (Gonzalez-Benito and Gonzalez-Benito, 2005).

Triple bottom line: social performance measurement

Corporate social responsibility (CSR) is often used to describe a firm's social activities. However, CSR means many different things to different people, and for this reason the social aspects of the triple bottom line are less understood and consistent than measures associated with environmental performance. A firm's social performance could include a host of organizational activities, including the donations of individual employees, or CSR could entail broad conceptualizations such as "corporate citizenship" or "strategic philanthropy." Until managers have commonly accepted a means of assessing social performance, assessment should include a combination of internally and externally focused measures. Employee well-being programs, implementing sourcing and vendor standards, and standardizing the firm's donations, sponsorships, and community outreach activities could also be included. Unfortunately, however, today there is no single widely accepted standard for social management systems to parallel the environmental performance measurements found in the ISO certifications.

Balanced scorecard

An alternative to the triple bottom line approach to assessing sustainable performance is to include social and environmental issues in the balanced scorecard. This is a widely recognized way to assess organizational performance, and lends itself to sustainable performance metrics because it already incorporates the perspectives of internal and external stakeholders.

Balanced scorecard: measurement

There are three alternatives for the inclusion of sustainability in the balanced scorecard.

- Integrate social and environmental measures into the existing matrix. For example, energy efficiency and water usage could be assessed as internal business processes.
- Rather than integrate sustainability targets into the existing scorecard, develop a separate sustainability scorecard. For example, it could be devoted solely to the firm's efforts towards reducing energy and waste, improving employee well-being, and community involvement.
- Add non-market factors to the scorecard. Environmental and social measures could be added as their own unique "quadrants," or "spokes on the performance wheel."

Balanced scorecard: environmental performance measurement

Whether adding to an existing scorecard or developing new quadrants, there are a handful of environmental and social measures that leaders should consider implementing.

- The efficient use of materials is an issue for all organizations. Leaders could begin by simply tracking the energy use per unit and water use per unit, as these are regarded by scientists working on environment issues as areas where organizations must reduce usage.
- Emissions is another area that must be reduced. However, for each organization, the specific type of emission would vary.

Balanced scorecard: social performance measurement

For social performance, leaders should consider choosing one measure for employees, one for suppliers (upstream), one for community, and one for philanthropy (to reflect groups that the organization chooses to support, and the amount of this support).

Capturing sustainability's returns on investments

Sustainability is typically defined to include environmental stewardship, human health and equity issues, and the social implications of decisions. While the importance of these issues is widely recognized, organizations are challenged when they try to integrate sustainability into their investment and operating decisions. Initiatives in sustainability often require capital resources. To identify the most appropriate areas in which to make investments in sustainability, firms need to estimate the potential returns on investments. The following sections help to identify how sustainability's returns on investments can be captured.

Start incrementally with smaller projects

After performing social audits, it is not uncommon for firms to become aware of the wide range of opportunities available to them to reduce their use of energy and water in their production and service delivery processes, as well as ways to reduce waste production. However, it is important for firms to take modest and incremental steps towards sustainability, rather than tackling multiple projects at once. Small and modest "wins" early on can help build the necessary momentum, visibility, and legitimacy among mid-level managers by conveying that sustainability makes effective business sense and "works."

Survey the end-users of your products and services

Along with taking measured steps early on, it is important that firms start off in the right direction. Therefore, rather than taking a shotgun approach to sustainability, firms should begin by talking with users of their products and services to determine the sustainability initiatives that are most salient to them. Beginning in this way can help firms be positioned to measure the success of their sustainability efforts.

Reference the sustainability efforts of a wide array of firms, including those outside the industry

Research has shown that legitimacy is important for a wide range of organizational phenomena. Certainly, building legitimacy is important to convince managers of the relevance of, and in making the business case for, sustainability. One way to help build legitimacy is to convey the success prominent firms across a wide range of industries have had when they began making focused attempts to reduce their waste, energy, and water usage.

Publicize your organization's sustainability plans

Perception greatly influences how stakeholders view organizations. It is important to publicize your firm's progress in reducing its energy and water dependence.

> **Key principle**
>
> The triple bottom line and balanced scorecard are two approaches to capturing sustainable performance in a holistic fashion.

Sustainability measurement fundamentals

Once a firm has decided that sustainability is an important organizational priority, its next steps include determining what to measure, how to measure, and how best to report the results to stakeholders.

What should firms measure?

One of the most important factors in sustainability measurement and reporting is picking the right metrics to report. There is a growing number of ways to measure social and environmental performance. However it is important to keep your stakeholders in mind when coming up with what you are to measure. Put simply, choose the metrics that will matter most to your firm's stakeholders. Because metrics surrounding your firm's sustainability efforts can help foster innovation and growth, it is important to focus on continuous improvement as the primary driver for measuring sustainability, not simply a box-ticking exercise. If metrics don't add value, it is unlikely they will support continuous improvement, and they will eventually be discarded by managers because they are perceived to be tasks that only require time and effort to collect. Compliance with regulatory requirements is a key benchmark for environmental performance. However, firms should not simply do the bare minimum when it comes to sustainability. Most firms today compete in a highly competitive environment. Having the appropriate set of metrics in place is important for firms to gauge their success in meeting short- and long-term business objectives. Measuring performance with a focus on sustainability is not just a new responsibility, it can constitute new opportunities that companies can embrace to drive organizational value.

When determining what to include in their sustainability reports, many firms simply follow the Global Reporting Initiative (GRI; www.globalreporting.org) guidelines. The GRI framework includes a series of principles that firms can use to determine whether a certain piece of information merits inclusion. These principles are materiality, stakeholder inclusiveness, sustainability context, and completeness.

*   **Materiality** refers to information that reflects the firm's most significant impacts on society and the environment. In order to determine what is material, a firm should assess its "overall mission and competitive strategy, concerns expressed by stakeholders, broader social expectations, and the organization's influence on upstream and downstream entities."

- **Stakeholder inclusiveness** refers to the need for reports to respond to stakeholders' reasonable expectations and interests.
- **Sustainability context** reflects the notion that the overarching purpose of a sustainability report is to offer firms an avenue to report to stakeholders how their activities are helping to improve, or at least reduce the negative impact upon, the environment.
- **Completeness** refers to the desire that sustainability reports include all significant impact of the business in order to better equip stakeholders attempting to assess the firm's environmental and social performance during the reporting period.

Box 8.3 outlines important factors that firms should consider when building metrics around social and environmental performance.

Box 8.3 Keys to sustainability performance measurement

- Measure factors that add value to organizational decisions
- Measure environment and social performance in much the same way that others around the world are measuring their sustainability initiatives
- While there is growing standardization of sustainability metrics, sustainability measurement offers firms a potential way to differentiate themselves from their competitors
- Large firms should develop metrics that enable their environmental and social performance to be derived for all their business units

---

Utilizing sustainability principles

Since the introduction of the Ceres Principles in 1989, sustainability reporting has been a key mechanism by which firms that choose to adhere to sustainability codes of conduct can demonstrate their accountability to outsiders.. The number and popularity of codes of conduct has greatly increased during the past two decades. In recent years, sustainability reporting has broadened to include social impact indicators and information on governance, and the target audience has broadened beyond shareholders and employees to include capital providers.

ISO 14000

Firms are increasingly taking a broader look at how they measure environmental performance with the advent of the ISO 14000 Standard and Specification and its companion guidelines over the past fifteen years (www.iso.org). In addition, the ISO 14031 Guidelines on Environmental Performance Evaluation provide for the establishment of measurable and verifiable environmental performance indicators appropriate to any public or private enterprise.

The Global Reporting Initiative

The GRI (www.globalreporting.org) is leading the pack in terms of setting reporting standards. It was launched by Ceres and has become the *de facto* international standard for corporate reporting on environmental, social, and economic performance. The reporting framework is organized around three main areas, and includes seventy-nine individual sustainability performance indicators. The primary areas covered are economic, social, and environmental. Reports based on the GRI framework can be used to benchmark organizational performance with respect to laws, norms, codes, performance standards, and voluntary initiatives; demonstrate organizational commitment to sustainable development; and compare organizational performance over time. Over 1700 organizations used the GRI framework to support their sustainability in 2010. In addition to the framework that all organizations must use, there are a number of sector-specific supplements covering electric utilities, financial services, food processing, mining and metals, NGOs, airport operators, construction and real estate, event organizers, media, oil and gas, automotive, logistics and transportation, public agency, telecommunications, and apparel and footwear.

Green Seal certification

Green Seal (www.greenseal.org) started in the late 1980s and has become a well known household sustainable products certification program. Green Seal has over thirty standards and covers more than 190 different product and service areas. Green Seal utilizes research that includes life-cycle assessment. Importantly, Green Seal uses a sliding scale, based on the revenue of organizations, that determines the annual costs of maintaining the Green Seal. The sustainable product certification provided by Green Seal is a specific and targeted program. For example, categories such as paints, household cleaners, and occupancy sensors (among others) are included within the sustainability standards.

Leadership in Energy and Environmental Design

One of the most advanced and comprehensive sustainability certification programs is the US Green Building Council's LEED Sustainability Certification Program (www.usgbc.org/LEED). The LEED system is currently in version III, and covers the following general development types: new construction, existing buildings (operations and maintenance), homes, schools, neighborhood development, retail, healthcare, and commercial interiors. Considerable documentation is required in order to meet the LEED requirements for each of these categories. Project managers typically turn to LEED consultants to help give input during a project and help meet the requirements for a given new or existing building.

The Greenhouse Gas Protocol Initiative

The World Resources Institute and the World Business Council on Sustainable Development jointly released the Greenhouse Gas Protocol (www.ghgprotocol.org) in 1998, which have since become the standard protocol for measuring greenhouse gas emissions. When organizations, large or small, need to measure their carbon footprint, they turn to this widely used two-phase methodology. Most organizations measure their scope I and scope II greenhouse gas emissions using this protocol. Firms then proceed to the second phase whereby their greenhouse gas inventory is verified by a third party, who provides oversight that the protocols were followed correctly and that the reported estimate is within 5% of the true value.

Carbonfree Product Certification

CarbonFund's Product Certification program (www.carbonfund.org/products) allows firms to offset the greenhouse gas emissions created during their production processes by purchasing carbon offsets from CarbonFund's portfolio of renewable energy projects. For every product that is certified, CarbonFund calculates the carbon footprint and/or the life cycle of the product using a methodology similar to the Greenhouse Gas Protocol Initiative.

Ceres Principles

The Ceres Principles for environmentally sound business practices (www.ceres. org/about-us/our-history/ceres-principles) is a ten-point code of corporate environmental conduct that firms choose to publicly endorse as an environmental mission statement or ethic.

Equator Principles

The Equator Principles (www.equator-principles.com) serve as a benchmark for the financial industry to identify and manage social and environmental issues in project financing.

United Nations Global Compact

The UN Global Compact's Ten Principles (www.unglobalcompact.org/AboutThe GC/TheTenPrinciples/index.html) focus on human rights, labor, the environment, and anti-corruption. Companies voluntarily commit to respect and implement the principles and follow up with communications on progress.

Caux Roundtable Principles for Responsible Business

The Caux Roundtable (www.cauxroundtable.org) Principles for Responsible Business are intended to serve as a measurable standard for responsible business worldwide. They are a statement of aspirations developed by an international network of business leaders who envision capitalism with principles.

Principles for Responsible Management Education

The Principles for Responsible Management Education (www.unprme.org) ask management educators to incorporate into their management education curricula a set of six principles that recognize the roles and responsibilities of businesses in creating a socially and ecologically sustainable world through the practice of responsible corporate citizenship.

Private Voluntary Organization Standards

The PVO Standards (www.interaction.org/document/interaction-pvo-standards) is a set of principles to enhance the professional, ethical, and responsible conduct of private voluntary organization members. These principles are a prerequisite for membership with InterAction, the largest coalition of US-based international NGOs engaged in humanitarian efforts.

Global Sullivan Principles

The Global Sullivan Principles (www.thesullivanfoundation.org) is a voluntary code of conduct intended to "be a catalyst and compass for corporate responsibility and accountability." Businesses of all sizes and sectors are encouraged to be responsible to employees and communities while pursuing business objectives.

OECD Guidelines for Multinational Enterprises

The Organization for Economic Co-operation and Development's Guidelines for Multinational Enterprises (www.oecd.org) is a set of recommendations for voluntary principles and standards of ethical business conduct for multinational enterprises operating in or adhering to OECD.

Sustainability metrics

There are a number of ways to measure the sustainability of different operations. Box 8.4 highlights the most popular ways to capture energy use, pollutant emissions, greenhouse gas emissions, and water and material intensities.

Box 8.4 Sustainability intensity indicators

- Energy intensity: kJ (kilojoule) equivalent per unit of output
- Pollutant emissions: kg of pollutants from the process per unit of output
- Greenhouse gas emissions: kg of CO_2 equivalent from the process per unit of output
- Water intensity: liters of clean water per unit of output
- Materials intensity: kg of material wasted per unit of output

---

The Institute for Supply Management (www.ism.ws) publishes a broad list of metrics and performance criteria for sustainability and social responsibility initiatives. The document lists the metrics for nine social responsibility categories: community, diversity, and inclusiveness; supply base, diversity, and inclusiveness; workforce; environment; ethics and business conduct; financial responsibility; human rights; health and safety; and sustainability. Box 8.5 lists some of the metrics and decision criteria in found in the ISM environment and sustainability category.

Box 8.5 Institute for Supply Management – selected environment and sustainability metrics

Environment metrics

- Disposal and waste management policies and practices
- Water conservation and consumption
- Green house gas (GHG) footprint (aggregate CO_2 number)
- Paper and paper product consumption
- Packaging reduction initiatives
- Energy consumption (power, gas, electric)
- Buildings and construction (LEED Certification, Green Globes, Energy Star, etc.)
- Transportation and logistics management, including routing and consolidation, fleet management
- Travel policies and statistics (miles driven, miles flown, nights away from base, etc.)
- Education and communication initiatives
- Contact information for Chief Sustainability Officer publicly available.

Sustainability metrics

- Use of sustainability criteria in procurement decisions
- Processes in place to embed sustainability and social responsibility into supplier qualification and certification decisions

---

- Processes in place to embed sustainability and social responsibility into product design, redesign, and statements of work
- Developing processes to ensure understanding of sourcing and recycling decisions
- Development of relationships with key suppliers to gain access to protected information on chemical makeup of products being purchased
- Working with risk management and/or internally to develop, quantify and base decisions on financial and other risks related to nonconformance with or lack of support of sustainability and social responsibility initiatives

Water stewardship tools

Because there are no clear-cut standards for water disclosure, firms are relying on existing sustainability and emerging water-specific voluntary reporting frameworks for guidance, such as the GRI. There are some emerging best practices with respect to water management, including those in Box 8.6.

Box 8.6 Best practices with respect to water management

- Providing board or executive committee oversight of water use
- Reducing reliance on drinking water by treating, reusing, and recycling industrial process water and captured rainwater and stormwater
- Partnering with communities, water utilities, and other industrial users to better manage regional water resources and share in common solutions
- Establishing quantifiable targets for reductions in freshwater use and wastewater discharge
- Assessing supply chain risks and building awareness among suppliers

---

Open-source water risk-analysis tools

The Global Water Tool

This tool (www.wbcsd.org/web/watertool.htm) combines geographic mapping with global water availability data to allow firms to compare water use with validated water and sanitation availability information on a country and watershed basis. The Global Water Tool rates facilities on local water stress and generates key water indicators used by the GRI.

Table 8.1 Key water measures

Measure	Objective
Total water withdrawal by source: surface water (rivers, lakes, wetlands); groundwater; rainwater; wastewater; water utility supplies	Provides baseline for improvement
Total water discharge by quality and destination	Identifies opportunities to reduce environmental impacts and related costs
	Indicates priorities for reducing risk of fines for noncompliance and loss of license to operate
Percentage and total volume of water recycled and reused	Demonstrates and tracks efficiency and reductions of total water withdrawals and discharges
	Identifies opportunities for reducing water consumption, treatment, and disposal costs
Water efficiency or water-use ratio (water used per unit of production, revenue, square footage, or employee)	Provides perspective on performance adjusted for changes in production quantity
	Improves ability to benchmark across operations and industry
Percentage of facilities operating in water-stressed areas	Identifies locations subject to increased risks from water scarcity, supply disruption, higher costs, regulatory changes, conflicts with local stakeholders, and resulting reputation damage
	Guides decision-making on prioritizing improvements and operational planning
Quantifiable targets for reducing water use and wastewater	Drives improvement and tracks performance
	Increases commitment and accountability

Source: PricewaterhouseCoopers, 2011

GEMI Water Sustainability Planner

"Collecting the Drops: A Water Sustainability Planner" (www.gemi.org/waterplanner) is provided by the Global Environmental Management Initiative (GEMI). This planner offers online tools that help assess a firm's relationship to water and the local communities where it operates, identify related risks, and develop a business case for improvement.

Comprehensive water risk analysis has helped Dow Chemical to prioritize improvement projects and improve planning at manufacturing plants located in the most water-stressed areas. Dow used the Global Water Tool to identify seven priority sites out of a potential 160. Then, using the GEMI Water Sustainability Planner, Dow managers collected water data and conducted in-depth risk surveys at targeted sites, and generated risk-factor scores based on local conditions for the watershed, supply reliability and economics, social context, compliance, and efficiency (Dow Chemical, 2008)

Key principle

There are many principles and environmental metrics firms can use to measure sustainable performance. Choose the principles and metrics that are most relevant to your organization's stakeholders.

Sustainability principles: the future

Each of the sustainability standards and principles identified above can help improve the materiality, consistency, and accountability of firms interested in improving their sustainability performance and in improving the transparency of the firm's impact on society and the environment. These standards can be useful sources for identifying performance measures that can potentially improve stakeholder relations.

While it is important that firms adopt principles in measuring environmental and social performance, the challenge associated with measuring sustainability is that there is no multi-industry, agreed-upon standard for assessing sustainability-related accounting. There are many reasons for this, not least the differing perceptions that academics and industry professionals have of the sustainability of firms. There is considerable variance in the findings of research devoted to the measurement of cost accounting and the challenges associated with accounting for a direct causal connection between the internal operationalization and performance effects of sustainability.

Sustainability performance assessment: external issues

In addition to developing internal measurement systems to track and capture the organizational outcomes of sustainability, leaders can also see performance outcomes through external sources. Over the past ten years, we have seen a significant change in the number of organizations rating the sustainability efforts of firms. New ratings have appeared and others have disappeared, while some ratings organizations have merged and others realigned. This growth is explained in part by increased awareness and acceptance of sustainable practices by corporations and their stakeholders, which in turn has created a growing market for information on corporate sustainability performance. Various organizations have jumped in to meet the demand, tailoring ratings to provide specific views on sustainability performance or to market the ratings to different target audiences. The growth in ratings has also been abetted – perhaps ironically – by the increasing amount of sustainability information disclosed by companies, which provides the raw material for additional company ratings.

Reporting sustainability results

Sustainability reporting provides a range of benefits. Most importantly, reporting the results of your firm's sustainability initiatives helps firms to disclose publicly their social and environmentally responsible activities, and to convey to stakeholders how their performance in these areas matches up to their stated goals. When sustainability reporting is done effectively, stakeholders (investors, customers, employees) can make more informed decisions about their relationships with your firm. In addition, sustainability reporting can help the leadership better align strategic goals with the firm's sustainability initiatives.

Sustainability reporting is still voluntary today. However, more than 3000 companies worldwide, including more than two-thirds of the Fortune Global 500, issue sustainability reports.

The most common approach to reporting on sustainability performance has been to publish a sustainability report, either in conjunction with, or separately from, the company's annual report (Jones *et al.*, 2005; O'Dwyer and Owen, 2005). It is quite likely that, in the not-so-distant future, firms will be required to report, at least at a bare minimum, a certain minimum threshold of social and environmental performance metrics. Therefore it is in the best interests of firms to begin now, rather than wait for regulations to mandate their sustainability reporting activities. Sustainability reporting is likely to become a core business activity.

Research indicates that equity analysts increasingly consider sustainability practices when valuing and rating public companies. For example, a recent global Ernst & Young survey of 300 executives at large international firms revealed that 43% believed equity analysts consider factors related to climate change when valuing a company (Ernst & Young, 2010). Another study, by George Serafeim at Harvard and Ioannis Ioannou of the London Business School, revealed that equity analysts give higher ratings to companies with excellent CSR practices. These scholars based their findings on a survey of over 4100 publicly traded companies over a sixteen-year period (Ioannou and Serafeim, 2010).

Benefits of external sustainability assurance and scrutiny

Sustainability assurance is an objective and systematic way in which firms can assess their social and environmental initiatives, how they are currently working, and how they may be improved. It is quite likely that firms can leverage internal resources to help audit processes in these areas. However, once a firm's sustainability information satisfies internal assurance standards, the firm should strongly consider subjecting its performance to external assurance standards from third-party groups that have experience in this area.

There are several benefits associated with accurate and complete sustainability reports. One of the more important is that rating agencies may refer to sustainability-related information to assign a grade to a firm's debt (Ernst & Young, 2010). Hence accurate sustainability reporting could potentially help lower a company's costs of borrowing. Therefore it is in a firm's best interests to have a third-party entity audit their sustainability report to assess whether their performance goals are being captured and measured correctly.

There are two primary global assurance standards. The International Standard on Assurance Engagements 3000 (ISAE 3000) is the benchmark most often used by accountants as the basis for assurance of sustainability reports. The ISAE 3000 was developed by the International Auditing and Assurance Standards Board. Another global standard is the AA1000AS (2008), which was designed for use outside of the accounting profession. The AA1000 is a principle-based standard. Ernst & Young reports that using both ISAE 3000 and AA1000AS is a leading practice in sustainability reporting (Ernst & Young, 2010). They also advise that country-specific general assurance standards (including the American Institute of Certified Public Accountants' AT101 and the Canadian Institute of Chartered Accountants' Handbook section 5025) can be helpful in non-financial reporting in their respective countries. Box 8.7 summaries the benefits associated with external sustainability assurance.

Box 8.7 Benefits associated with external sustainability assurance

- Greater awareness and comprehension of sustainability performance
- Better identification of areas needing attention and improvement
- Enhanced legitimacy and reputation with both internal and external stakeholders
- Improved environmental and social controls and processes
- Increased awareness and connection of top managers with social and environmental initiatives and organizational outcomes
- Better decision-making that links sustainability initiatives across business units

---

Key principle

Using ISAE 3000 and AA1000AS is a leading practice in sustainability reporting.

Sustainability ratings and certifications

A recent empirical study analyzed the shareholder value effects of environmental performance by measuring the stock market reaction associated with announcements of environmental performance (Jacobs *et al.*, 2010). Their study focused on reviewing how the market values impacts of environmental announcements. Their sample was based on analysis of 811 announcements (430 Corporate Environmental Initiatives announcements and 381 Environmental Awards and Certifications announcements) that appeared in the daily business press during the period 2004–06. Their study's results uncovered a number of interesting findings, including that announcements of gifts to environmental causes are associated with significant positive market reaction; voluntary emissions reductions are associated with significant negative market reaction; and ISO 14001 certifications are associated with significant positive market reaction (ibid.). The authors also found that the market reacted positively (on a statistical basis) to announcements of ISO 14001 certifications. This is the first study to provide empirical evidence of the impact of ISO 14001 certification on market value (ibid.).

Evaluations based on more recent certification announcements yield similar results. A recent article in *Triple Pundit* (Covington, 2011) cites a recent *Fast Company* article concerning Bloomberg's business of measuring companies' "environmental, social, and governance" (ESG) performance, which found that "the number of investors accessing ESG data is up by 29% comparing the first half of 2010 with the second. Investors use it to identify smart practices – for example, companies who operate in a socially responsible manner may be viewed as forward thinking and well managed." Covington concludes that "this surely portends that markets will inevitably respond favorably to sustainability efforts, especially when the data shows improved governance and profits result directly, and in the long run, from sustainability" (ibid.).

There are other noteworthy studies that demonstrate the value external stakeholders place on sustainability certifications. For example, a report by the Arava Institute for Environmental Studies suggests that a supplier that was awarded the ISO 14001 or Eco-Management and Audit Scheme (EMAS) certification by an independent entity enhances perceived reliability and is demonstrating ethical responsibility (Bellesi *et al.*, 2005). Moreover, the importers evaluated felt more confident engaging a new supplier when they had ISO 14001 certification in that it helped them save time and effort, and reduced uncertainties in the periods leading up to their purchase orders. Like other forms of certifications by credible third-party groups, these examples underscore how ISO 14001 accreditation confers economic benefits and greater "market value." Another study found significant differences in the market value of firms with ISO14001 certification and those without (Murillo-Luna and Ramon-Solans-Prat, 2008). In addition to the external benefits, internal efficiencies are considered significantly higher for firms with ISO14001 certification. Hence investments in ISO14001 certification can lead to both internal and external benefits for firms.

Are all sustainability raters the same?

In our previous sections we have highlighted how important sustainability performance has become to a wide array of organizational stakeholders, from asset managers and employees, through suppliers, to citizens and consumers. With this increased attention has come awareness that sustainability ratings should be able to demonstrate that they are fair, accurate, and credible. A recent series of reports set out to identify ratings organizations that exhibit strong practices in select areas and to provide a forum through which the various stakeholders in the ratings arena – raters, companies, investors, sustainability experts, etc. – can learn from these practices and share perspectives on how ratings need to evolve going forward to ensure they are credible and robust (SustainAbility, 2011). Their findings are quite instructive and are summarized below.

- There are a growing number of ratings that cover specific issues, industries, and regions. However, ratings that cover multiple issues, industries, and/or regions are predominant.
- Most raters base their rankings on public information, therefore firms benefit from being timely with their sustainability disclosures.
- Companies that do not respond requests from rating organizations for sustainability-related information typically fare worse than responders.

Rating agencies collect their data in various ways, such as studying CSR reports, administering voluntary corporate surveys, analyzing media reports, engaging in independent investigations, and actively communicating with the management of rated corporations. The ratings most often produce a quantitative analysis, although some offer narrative discussion, exclusively or in addition to the quantitative results. Some ratings are based solely on non-financial data to assess social responsibility independently from financial performance. Others combine financial and non-financial data to measure long-term value and sustainability. All the rating agencies have a goal of providing some objective and verifiable grounds for assessing what is, at least in part, the non-financial social performance of a corporation.

An overview of prominent sustainability raters

Ratings can help spur constructive conversations about what constitutes good social and environmental performance. Certainly, ratings organizations were once the domain of niche organizations. However, sustainability ratings have become mainstream, with players such as Goldman Sachs (GS Sustain Focus List), *Newsweek* (Green Rankings), Thomson Reuters (ASSET4), Walmart (Sustainability Index), and Standard & Poor's (S&P/IFCI Carbon Efficient Index, S&P ESG India Index) now attempting to measure sustainability performance. A complete list of third-party sustainability rating organizations is available in Appendix B.

Key principle

Frequent reporting and transparency are keys to being evaluated highly by external sustainability rating groups.

Summary

- Pressures are continually mounting for firms to initiate sustainability initiatives and pursue efforts to improve their social and environmental performance records.
- Unfortunately, all too often managers do not fully recognize the performance implications of sustainability
- The performance implications of sustainability can be seen not only at the broadest corporate level, but also at the level of the individual business unit, and through endorsement and certification by influential third parties.
- When choosing sustainability performance metrics, firms should consider those that are most salient to their stakeholders.
- Firms should report their environmental and social performance in a timely fashion.

Discussion questions

- What are the benefits that firms can derive through sustainability? Identify both intangible and tangible benefits.
- Identify how firms can assess their sustainability performance.
- Compare and contrast the leading sustainability principle organizations.
- What are the leading ways to measure firms' water usage?
- Who are the leading third-party sustainability rating groups today?
- What are emerging trends among sustainability raters, and among firms that are rated highly for their environmental and social performance?

References

Bellesi, F., Lehrer, D., and Tal, A. (2005) "Comparative advantage: the impact of ISO 14001 environmental certification on exports," *Environmental Science & Technology*, 39: 1943–1953.

Busco, C., Frigo, M., Leone, E., and Riccaboni, A. (2010) "Cleaning up: implementing sustainability by using management," *Strategic Finance*, 92: 29–37.

CIM (2007) *Shape the Agenda: The Good, the Bad and the Indifferent – Marketing and the Triple Bottom Line*, Chartered Institute of Marketing, Maidenhead, UK, www.oisolutions.co.uk/cms/site/docs/Triple_Bottom_Line_Agenda_Paper.pdf

Clorox (2010a) "Green Works® Natural Cleaners and Sierra Club® Celebrate Two-Year Anniversary; Doubling of Natural Cleaning Category," The Clorox Company, Oakland, CA, www.reuters.com/article/2010/03/01/idUS177189+01-Mar-2010+BW20100301

——(2010b) "Corporate social responsibility," www.cloroxcsr.com

Covington, P. (2011) "Sustainability's influence on stock value," *Triple Pundit*, www.triplepundit.com/2011/03/stock-market-view-sustainability

Dow Chemical (2008) *2008 Global Reporting Initiative Report*, Dow Chemical, Midland, MI, www.dow.com/sustainability/pdf/GRI_71409.pdf

Ernst & Young (2010) *Climate Change and Sustainability: Seven Questions CEOs and Boards Should Ask About 'Triple Bottom Line' Reporting*, Ernst & Young, London, www.ey.com/climatechange

Fromartz, S. (2009) "The mini-cases: 5 companies, 5 strategies, 5 transformations," *MITSloan Management Review*, 51: 41–45.

Gonzalez-Benito, J. and Gonzalez-Benito, O. (2005) "An analysis of the relationship between environmental motivations and ISO14001 certification," *British Journal of Management*, 16: 133–148.

Hespenheide, E., DeRose, J., Branhall, J., and Tumiski, M. (2010) "A profitable shade of green," *Deloitte Review*, 7: 61–73.

Hewlett-Packard (2009a) *2009 HP Global Citizenship Report*, Hewlett-Packard, Palo Alto, CA, www.hp.com/hpinfo/globalcitizenship

——(2009b) "Eco-friendly data centre promises windfall For HP clients," Enterprise Services, Hewlett-Packard, Palo Alto, CA, http://h30423.www3.hp.com/?rf=sitemap &fr_story=4fde9b9dab96664b74e2f13552bf28aa99ed2e9c&jumpid=reg_R1002_USEN

——(2010) "The HP printing payback guarantee," Hewlett-Packard, Palo Alto, CA, www.hp.com/large/campaign/guarantee/index.html

Ioannou, I. and Serafeim, G. (2010) *The Impact of Corporate Social Responsibility on Investment Recommendations*, Working Paper No. 1507874, Harvard Business School Accounting & Management Unit, Cambridge, MA, http://ssrn.com/abstract=1507874

Jacobs, B. W., Singhal, V.R., and Subramanian, R. (2010) "An empirical investigation of environmental performance and the market value of the firm," *Journal of Operations Management*, 28(5): 430–441.

Jones, S., Frost, G., Loftus, J., and Van der Laan, S. (2005) *Sustainability Reporting: Practices, Performance and Potential*, CPA Australia, Sydney, www.cpaaustralia.com. au/cps/rde/xbcr/cpa-site/sustainability-reporting-practices-performance-potential.pdf

Murillo-Luna, J. and Ramon-Solans-Prat, J. (2008) "Which competitive advantages can firms really obtain from ISO14001 certification?," *Journal of Industrial Engineering and Management*, 1: 104–108.

Nike (2010) *CR Strategy Overview, Corporate Responsibility Report FY07–09*, www.nikebiz.com

O'Dwyer, B. and Owen, D. (2005) "Assurance statement practice in environmental, social and sustainability reporting: a critical evaluation," *British Accounting Review*, 37: 205–229.

Potier, B. (2002) "Treasury secretary talks values at HBS," *Harvard University Gazette*, October 24, http://news.harvard.edu/gazette/2002/10.24/13-oneill.html

Reinhardt, F. (1999) "Bringing the environment down to Earth," *Harvard Business Review*, 77(4): 149–157.

SEC (2010) "Interpretive Guidance on Disclosure Related to Business or Legal Developments Regarding Climate Change," US Securities and Exchange Commission, Washington, DC, www.sec.gov/rules/interp/2010/33-9106.pdf

SustainAbility (2011) *Rate the Raters, Phase Three: Uncovering Best Practices*, SustainAbility, www.sustainability.com/library/rate-the-raters-phase-three

9   Summary

Throughout *Leading the Sustainable Organization* we have presented the components of an integrated model developed to help leaders in any organization design, implement, and assess their sustainability efforts by engaging their workforce at all levels. The framework presented addresses macro-level (organization-wide) as well as micro-level (manager-to-employee) leadership and employee engagement facets of sustainability. At the macro-level, leaders can use our model as a road map to develop and reinforce a clear sustainability direction for their firms. And at the micro-level, leaders can use the model to help them build into their organizations manager-to-employee practices that individualize the employee engagement required to implement sustainability initiatives effectively across their organizations.

The Leading the Sustainable Organization model is applicable to any size of organization and to any industry, and applies to all firms no matter what their current stage of sustainability development. Likewise, researchers can use the model to frame single-discipline or single-level enquiries within the broader context of multi-faceted sustainability efforts. Not only is the model grounded in sound leadership and human capital-related theories, but throughout the book, practical examples are provided of how each element of the framework is being implemented, from large and small-to-medium sized firms from various industries, including National Geographic, Mission Foods, Four Seasons Hotels, Marks & Spencer, Bridgestone, Nike, Bristol-Myers Squibb, P&G, Siemens, and others.

> **Key principle**
>
> The Leading the Sustainable Organization model is based on sound theory combined with examples of practical application from numerous organizations across various industries.

Alignment is essential

Each component of the Leading the Sustainable Organization model adds value to an organization's pursuit of sustainability. However, taken individually, each component will not suffice to help management develop or reinforce a firm's sustainability efforts. For example, building sustainability into a firm's mission and values are essential components of setting a company's sustainability direction. Yet, whatever good work might be accomplished to create a firm's sustainability and values will be lost without the sustainability direction being reinforced through the other elements of the model. Therefore aligning each element of the framework with all the other elements is a key aspect of success. At the macro-level, the alignment of a firm's mission, values, goals, strategy, and HR value chain is crucial. Moreover, the macro-level factors must also be aligned with the micro-level factors: individual employee goals, job design, and transformational leadership. Doing so builds and reinforces a firm's sustainability agenda throughout the organization, at all levels.

Key principle

Alignment of each component of the Leading the Sustainable Organization model with the other elements of the framework is crucial to success.

Language matters

There is an old parable that tells the story of three stone workers at a construction site. All were doing the same job, but when each was asked what his job was, their answers differed: "Breaking rocks," replied the first; "Earning a living," answered the second; "Helping build a cathedral," responded the third. The same parable can easily transfer to corporate sustainability efforts. When asked what is their role in the firm's sustainability efforts, three different employees can easily give three very different answers. For example, "Saving the company money," replies the first; "Saving the company resources," answers the second; "Helping future generations," responds the third.

Understanding how language works is essential to understanding how organizations work. An appreciation and understanding of language is important not only to researchers and consultants working with organizations, but also to the people working in them (Musson and Cohen, 1999). Boden (1994: 202) argues that organizations are constantly created and recreated through the "unfolding dynamism of talk." Likewise, Boyce (1995) explored the role of organizational stories in creating a sense of unity and common purpose among organizational members, concluding that shared language expresses shared experience and can be used to create a collective awareness among the workforce, as well as to change the organization. The research of Pondy (1978), Conger (1991), and Bate (1994) found that language skills are essentially power skills in that they hold the basis of influence within organizations. Therefore the persuasive aspect of language is a key leadership skill. This is especially important within today's organizations, when persuading people to adapt to constant change is a central leadership role (Wilson, 1992). Furthermore, as Conger (1991: 31) points out, because of the requirement that organizations need to be flexible and responsive to ever-changing environments, "the era of managing by dictate is ending." Consequently, the ability to motivate, persuade, and assure within an organization becomes even more essential to effective leadership. Musson and Cohen (1999: 35) affirm that "The ability to translate ideas into action, to interpret between the different worldviews within and outside the organization, to create a framework for action which everybody can understand and accept is central to the craft of leadership." Fundamental to this capability is the art of expression through which a leader becomes a meaning-maker, translating new concepts into a new kind of "common sense" (Bate, 1994: 257) upon which the organization can act.

Because of the power of language in organizations, leaders must continually be conscious of the language they use to portray their sustainability efforts. First, leaders should use consistent language to describe such initiatives. Consistency of language to express the sustainability agenda of the firm helps promote a consistent understanding among the workforce of how the firm defines sustainability. Likewise, language consistency facilitates alignment across the organization (between departments, functions, and geographies), as well as among organizational levels (from executives, to managers, to front-line employees) about why the firm is pursuing sustainability and how various areas of the firm can participate in the effort.

In addition to consistency, positive language is imperative to communicate commitment on the part of leadership to the firm's sustainability direction. In contrast, avoiding negative language discourages skepticism among the firm's workforce. Here are several common "killer phrases" of which management should be aware, and which should be avoided.

"Sustainability is easy"

Sustainability can often be seen as something that is simple and straightforward. By making this assumption, leaders can underestimate the effort a well thought-out, long-term sustainability program will take. For example, a firm can put out recycling bins in each location or tell employees to turn off the lights when they leave a room. The reality is that well thought-out sustainability efforts are multifaceted and touch virtually every aspect of the firm. Communicating to the workforce that there are some simple things the organization can do around sustainability is a good start. But letting everyone know that their ideas for creating a more sustainable organization are welcome, and then acting upon those ideas, is vital.

"Sustainability is management's responsibility"

As discussed in previous chapters, leaders definitely have a role in sustainability efforts. However, because comprehensive sustainability efforts touch every part of the organization at all levels, employees throughout the organization must be engaged in helping make the firm's sustainability agenda successful.

"Sustainability can wait, we have more important things to do right now"

Taking this attitude toward sustainability simply delays the inevitable. As sustainability efforts are undertaken by a firm's competitors – saving them money, improving their brand with consumers, and making them a more attractive employer to job-seekers – the firm that puts off sustainability will have to play catch-up simply to survive.

"We don't have the resources to devote to a sustainability effort right now"

A key component of sustainability initiatives is that they ultimately save resources. Although people across the organization need to devote some time and effort to making sustainability work for the firm, the net gains in resource savings (money and materials) can be considerable.

"Sustainability is for tree-huggers"

It may be true that sustainability efforts save natural resources, which would be used more quickly and in greater quantities if firms continued operating in their current status. However, saving money and resources, becoming more attractive in the eyes of customers, and turning into a more competitive employer are sustainability results for firms that any manager and employee can appreciate.

Key principle

The language leaders use to talk about the sustainability agenda of the firm has a profound impact on the attitudes and behaviors of the workforce regarding the firm's sustainability efforts.

The seven deadly sins of sustainability efforts

1. Assuming everyone defines sustainability the same way

As discussed earlier in this book, sustainability can mean different things to different people. Throughout the literature, various definitions of sustainability exist. Sustainability, corporate social responsibility (CSR), "going green," and the "triple bottom line" are all common phrases that are used to refer to the phenomenon. When embarking upon sustainability efforts, leadership must first ensure that a consistent definition is agreed upon, which can be communicated to the broader organization and its stakeholders. As a reminder, the definition we have used in this text is that of the World Business Council for Sustainable Development: meeting the needs of the present without compromising the ability of future generations to meet their own needs.

2.   Not knowing why the organization is pursing sustainability

Once defined, different people can hold different opinions about the reasons a firm should pursue sustainability, and they may even question whether the firm should pursue it at all. For example, some people see corporate sustainability as essential for cost savings or risk avoidance; other people view it as a necessity for society as a whole; others believe companies are responsible for adding value to stakeholders beyond shareholder value; some see it as a competitive differentiator; while other people don't see sustainability as being important at all, with total disregard for any impact firms have beyond their requirement to make a profit. Because people have varied opinions about the value of sustainability and the reasons for pursuing it, leaders must ensure the reasons why their firm is pursuing sustainability are clear to everyone in the organization.

3.   Only seeing the big picture

Another mistake leaders often make is to set a sustainability direction for the firm at the macro-level, but to ignore the micro-level. Leaders must remember that the macro elements (mission, values, goals, strategy) only set the direction, and the HR value-chain processes serve to reinforce the firm's sustainability agenda at the organization-wide level. In order for sustainability to take root and develop across the organization, employees at all levels must be engaged in the effort through the micro-level (manager-to-employee) leadership practices discussed earlier.

4.   Focusing only on the operations and ignoring the people

Sustainability is most tangible in the operational changes that are made to save money and resources. Energy savings, reduced water use, waste disposal, and recycling, for example, are all clearly seen and are measurable. However, these changes do not happen by themselves. They must be identified, designed, and implemented by the firm's workforce. Engaged employees are much more likely to help recognize and execute sustainability initiatives than those who are not engaged. Moreover, engaged employees are much more likely to participate in the social aspects of sustainability through volunteer efforts both inside and outside the firm; and by putting in effort above and beyond their immediate job description.

5.   Not measuring the outcomes

As discussed previously, sustainability efforts can and should be measured. Assessing progress against macro-level, firm-wide sustainability goals is crucial, for several reasons. First, measurement enables the firm to assess progress against established sustainability goals. Second, measurement allows the organization to determine whether or not the firm-wide sustainability actions are having an effect and, if not, measurement can help identify the reasons for the lack of impact. For example, the firm might be taking the wrong action; or the actions taken might be useful, but more may need to be done to achieve the sustainability goals.

At the micro-level, measuring employee-level sustainability actions is equally as important as measurement at the firm-wide level. Employee-level measures serve to assess progress against individual sustainability goals. Like organization-wide measures, employee measures can help identify whether, and to what extent, each individual is engaged in the firm's sustainability efforts.

6. Seeing sustainability as a fad

Although popular management approaches (such as total quality, re-engineering, delayering, and matrix management) seem to come and go, sustainability is nothing less than an imperative for the future of society as a whole. Resources, including energy, food, water, and other raw materials, are in greater and greater demand. Reasons for this increased demand, with no decrease in the foreseeable future, include the continuous growth of the global population (achieving the 7 billion milestone as of 2011); the expansion of emerging market producers such as China, India, Brazil, and many others; and the desire for consumers around the world to accumulate possessions that once were reserved only for more developed economies. Without sustainability efforts by corporations and individuals alike, the strain on natural resources will continue to grow at ever-increasing rates. Sustainability may not, however, eliminate or even decrease demand for resources. At best, sustainability efforts across multiple organizations around the globe may only slow demand down.

7. Expecting perfection from day one

All too frequently, people are quick to criticize initiatives that are not a success from day one. However, any effort that involves the scope, scale, and duration of sustainability cannot be perfect from the outset. Seasoned leaders have learned that the successful execution of any large-scale organizational effort (mergers, acquisitions, new market entry, product launches, and so forth) requires multiple adjustments along the way. Too many variables exist across organizations for sustainability efforts to be absolutely predictable in their implementation and outcomes. Therefore leaders must view effective sustainability initiatives as an ongoing process of continuous learning (from both successes and mistakes) and improvement.

Key principle

Committing any one of the seven deadly sins of sustainability will damage the organization's efforts, and committing multiple transgressions will ensure frustration and potential failure.

Key success factors

Sustainability is a marathon, not a sprint

There are many sustainability "quick wins" that firms can achieve right out of the gate. For example, providing recycling bins and installing automatic on/off light switches will yield savings and resource reduction right away. However, to truly live by an organization's sustainability mission and values, and to achieve annual sustainability goals and strategies, a firm must view its sustainability agenda as entailing short-term (30–90 days), medium-term (90–180 days), and long-term (180–365 days and beyond) efforts. To achieve far-reaching sustainability goals, leaders must regard sustainability as a multi-year endeavor.

Sustainability is everyone's job

Sustainability is not just management's job, or the job of a single function. It cannot be assigned to a special "sustainability team." To be successful, sustainability has to be seen as everyone's job. Executives, managers, and front-line employees, across all locations and geographies, need to become engaged in the effort if broad sustainability goals are to be realized.

Think globally

Sustainability is a systemic issue. To achieve the sustainability gains available to a firm, leaders must have the capacity to see the interconnectedness between various components of their organization (functions and geographies), and between their firm's internal operations and the external communities in which they operate. Leaders must also be able to communicate a broader, global view to their management and employees in order to encourage them to think systemically about the business and how they can contribute to improving its "triple bottom line."

Every little bit helps

An easy position to take about sustainability is that one person, or even one firm, cannot make a difference in the large scheme of the global market. It may be true that the resource savings and social improvements that one person or one organization can make are minute compared with the scale of the global economy and world population. However, everything counts in large amounts. Each day, billions of people around the world go to work and use energy, water, and other natural resources that collectively add up to an enormous impact on a global scale. In the same regard, when those same people collectively engage as a global workforce in multiple sustainability efforts, the scale of their resource savings is also enormous.

Inspire others to act

Organizations that are successful in their sustainability endeavors set examples for others to follow. As an organization reduces costs, builds its brand, and becomes a more attractive employer due to its sustainability achievements, other firms not only want to imitate those achievements, but in order to compete they have to accomplish similar results. Similarly, when individual employees are recognized for their sustainability achievements, other employees are also inspired to act.

Key principle

Taking a long-term, systemic perspective and involving all employees enables a firm to become a sustainability leader.

Your personal approach to sustainability

Beyond the leadership practices described throughout this text, leaders are also individuals. Before you can set a sustainability direction for your firm, build HR value chain practices that reinforce that direction, and inspire others to act on the organization's sustainability efforts, you should understand your own personal approach to sustainability. Without embracing an obligation to the positive societal impact at a broad level, or, at a minimum, committing to the operational and financial gains that can be achieved, you will marginalize your firm's pursuit of sustainability from the outset. Therefore you should conduct a "sustainability self-assessment" (see the Key tool below, "My personal sustainability approach at work") that includes:

- determining whether you value sustainability as a personal and organizational endeavor;
- clarifying why you are considering incorporating sustainability into the firm's business model;
- understanding how you communicate with others about sustainability;
- determining whether you are willing to make sustainability part of your own personal performance plan;
- identifying how you can demonstrate to others your commitment to pursuing sustainability;
- understanding how you can incorporate sustainability into your decision-making; and
- determining your willingness to participate in sustainability activities above and beyond your day-to-day work.

Key principle

Understanding your personal approach to sustainability will help you to be a more effective leader of your firm's sustainability efforts.

Summary

- Throughout *Leading the Sustainable Organization*, we have presented the components of an integrated model developed to help leaders in any organization to design, implement, and assess their sustainability efforts by engaging their workforce at all levels.
- The Leading the Sustainable Organization model is applicable to any size of organization and to any industry, and applies to firms no matter what is their current stage of sustainability development.
- Aligning each element of the Leading the Sustainable Organization model with all of the other elements in the framework is a key aspect of success.
- Because of the power of language in organizations, leaders must continually be conscious of the language they use to portray their sustainability efforts.
- Committing any one of the seven deadly sins of sustainability will damage the organization's efforts, and committing multiple transgressions will ensure frustration and potential failure.
- Taking a long-term, systemic perspective and involving all employees enables a firm to become a sustainability leader.
- Conducting a "sustainability self-assessment" will help you to lead and to inspire others to act on the organization's sustainability efforts.

Discussion questions

- Which components of the Leading the Sustainable Organization model is your firm implementing well? Which components need to be improved?
- Within your firm, is each component of the Leading the Sustainable Organization model aligned with all the other components?
- What language do you use to communicate with others about sustainability? What language do you hear other leaders in your organization using? How can you, and the other leaders in your organization, improve how you talk about sustainability?
- Which, if any, of the "seven deadly sins of sustainability" is your organization guilty of committing? How might those transgressions be rectified?
- Is your organization following the "Key success factors" described above? If so, how? If not, how can the organization do a better job of using the key success factors?
- What is your personal approach to sustainability? What do you feel you do well in your personal approach? What aspects of your personal approach do you feel you can improve upon?

Key tool

My personal sustainability approach at work

Completing the following scorecard will provide a quick view of the degree to which you are integrating sustainability into your day to day work.

Steps to complete the assessment

1. Rate each item on a scale of 0 (poor) to 10 (excellent).
2. Make notes for each item to explain the rationale for the numerical rating.
3. Add all ten scores to obtain a total score (maximum = 100).

Rating scale

- 0–20 = poor (significant improvement needed across most or all components)
- 21–40 = below average (improvement needed in several components)
- 41–60 = average (identify areas of weakness and adjust)
- 61–80 = above average (identify areas that can still be improved)
- 81–100 = excellent (continuously review and refine each component as the firm's sustainability efforts evolve)

Table 9.1 My personal sustainability approach to work

Component	Rating (0 = poor; 10 = excellent)	Notes/ rationale
1 I value sustainability (e.g. I think sustainability is important for everyone to pursue in their day-to-day work)		
2 I know the firm's sustainability mission, values, goals, and strategy		
3 I regularly communicate in a positive way with people at work about sustainability		
4 I seek out and attend sustainability-related training and education at work, or outside work		
5 I build sustainability into my performance management plan (goals and measures)		
6 I consider sustainability in my decision-making at work		
7 I offer my sustainability ideas to the firm's leaders		
8 I participate in sustainability activities above and beyond my day-to-day work (e.g. volunteering)		

Component	Rating *(0 = poor; 10 = excellent)*	Notes/ *rationale*
9 I stay at my current company because the firm's sustainability values are a good match for my sustainability values		
10 I feel I am engaged in my current company's sustainability efforts		
Total score		

References

Bate, P. (1994) *Strategies for Cultural Change*, Butterworth-Heinemann, Oxford, UK.

Boden, D. (1994) *The Business of Talk*, Blackwell, Cambridge, UK.

Boyce, M.E. (1995) "Collective centering and collective sense-making in the stories and storytelling of one organization," *Organization Studies*, 16(1): 107–137.

Conger, J.A. (1991) "Inspiring others: the language of leadership," *Academy of Management Executive*, 5(1): 31–45.

Musson, G. and Cohen, L. (1999) "Understanding language processes: a neglected skill in the management curriculum," *Management Learning*, 30(1): 27–42.

Pondy, L. (1978) "Leadership is a language game', in M.W. McCall and M.M. Lombardo (eds) *Leadership: Where Else Can We Go?*, Duke University Press, Durham, NC.

Wilson, D. (1992) *A Strategy of Change: Concepts and Controversies in the Management of Change*, Routledge, London.

Glossary of key terms

Community citizenship behavior (CCB):   Discretionary actions of employees that are above and beyond their job description, occurring outside of the firm, aimed at bettering the community in which employees live and work.

Corporate social performance: The combination of a firm's principles of social responsibility and its policies, programs, and observable outcomes relating to the firm's societal impact.

Corporate social responsibility (CSR):   A company's sense of responsibility towards the community and environment (both ecological and social) in which it operates.

Employee engagement: An individual's involvement and satisfaction with, as well as enthusiasm for, work.

Extra-role performance: Employee behaviors, which occur within the firm, that are discretionary and exceed the requirements of their job description.

Full-range leadership: A combination of transactional and transformational leadership.

Going green: A firm's transformation efforts with the aim of becoming more environmentally sustainable.

HR value chain: The organization-wide systems and processes to attract, retain, engage, and develop a workforce that possesses values and talents that are aligned with the firm's mission, values, goals, and strategy.

In-role performance: The achievement of those tasks that are explicitly identified in an employee's position description and evaluated in the performance appraisal process.

Mission: Identifies how a firm defines itself and establishes the priorities of the organization.

Non-governmental organization (NGO):   A legally constituted entity, created with no participation or representation of any government.

Organizational citizenship behavior (OCB):   The behavior of an employee that is discretionary, not rewarded or recognized in an explicit way by the organization, and tends to promote efficient and effective functioning of the organization.

Strategy: The adoption of courses of action and the allocation of resources necessary for achieving organizational goals.

Sustainability: Meeting the needs of the present without compromising the ability of future generations to meet their own needs.

Triple bottom line: The financial, social, and environmental effects of a firm's policies and actions that determine its viability as a sustainable organization.

Transformational leadership: A leadership style that includes one or more of the following behaviors: idealized vision, inspirational motivation, intellectual stimulation, and individualized consideration. These leadership behaviors transform followers and motivate them to transcend their self-interests for the good of the organization.

Transactional leadership: A process that emphasizes the transactions and exchanges that take place between leaders and their followers. These exchanges are based on the leader identifying performance requirements and clarifying the conditions under which rewards are available for meeting these requirements.

Values: Beliefs about the types of goals firm members should pursue, as well as ideas regarding standards of behavior organizational members should use to achieve these goals.

Appendices

Appendix A – Executive summary of the "Current state of sustainability leadership" survey

Respondents

A total of 124 employees and managers from seventeen different industries responded. A majority (58%) of respondents held some level of supervisory or management position (front-line supervisor 16%, middle manager 33%, executive 9%) in their company. The majority worked in information technology (31%), followed by finance (20%), sales/marketing (19%), and manufacturing/ distribution/service delivery (14%). The largest industry groups represented (in descending order) were: financial services, information technology, manufacturing, telecommunications, consumer packaged goods, and healthcare. Industries represented were:

- consumer packaged goods
- education
- financial services
- food services
- healthcare
- information technology
- insurance
- manufacturing
- media
- professional services
- retail
- social services
- telecommunications
- transportation
- travel
- utilities
- wholesale distribution
- other

Results

The results of the survey indicate that the firms in our sample do in fact see sustainability as a means to achieve competitive advantage. The majority of our respondents (58%) indicate that sustainability ha⁻ entered into the lexicon of the "mission" of these firms. Additionally, 61% ⌐ported that sustainability was part of the firm's strategy, and 59% indicated that sustainability was part of the organization's values.

The organizations in our sample appear to be pursuing sustainability for a variety of reasons. The strongest reason (35% of responses) appears to be a perception that sustainability can provide a competitive advantage. The second strongest purpose for pursuing sustainability was risk management, or the avoidance of fines. Many firms believe that sustainability is the right thing to do for society (13%), and 8% believe that sustainability is the responsibility of all firms.

While these results indicate a strategic commitment to sustainability, that commitment does not appear to have been translated into an operational commitment at this time. It seems that the sustainability initiatives in these organizations are primarily focused outwardly, and have yet to establish a strong internal focus. These results suggest that sustainability is currently seen as a means of achieving external legitimacy and external resources, not as a means to develop and retain internal resources.

Despite the firm-level appreciation and support for sustainability, sustainability has not been translated into departmental goals and individual employee performance objectives. Consequently, employees may not feel that sustainability is "real." Until the sustainability strategies are translated into operational strategies and performance plans, the sustainability strategies of these firms will be hindered.

We have advocated the use of goal setting and performance management as important ingredients in engaging employees in their organization's sustainability efforts. When done properly, the sustainability mission cascades throughout the organization and is represented in the performance objectives of departments and goals of individual employees. At this stage of the sustainability movement, this level of involvement has not yet been achieved. In our sample, only 33% of respondents reported that they had specific sustainability-related objectives as part of their individual performance plan. Even fewer, 24%, reported receiving regular feedback about their efforts to implement sustainability practices in their jobs.

Despite the lack of systematic implementation, many of our respondents indicated a personal commitment to sustainability. In fact, 64% of those responding to our survey indicated that they engage in activities on their own, not with the aid or guidance of managers.

Taken as a whole, the results of our survey indicate that the sustainability initiatives in the firms represented in our sample are immature. Leading these organizations to a mature state of sustainability requires implementation of the practices identified in the Leading the Sustainable Organization model (Figure 1.1). Leading a sustainable organization requires a continuous focus that addresses both the macro (organization-wide) and micro (manager-to-employee) issues identified throughout this book. An organization's commitment to sustainability must transcend a merely external focus and permeate the entire organization by engaging individual employees in this effort.

Appendix B – Sustainability raters

The following list provides, in alphabetical order, an overview of various public and private entities that provide sustainability ratings and/or rankings of organizations.

Access to Medicines Index

www.accesstomedicineindex.org
    The Access to Medicine Index is a ranking of the world´s largest pharmaceutical companies on their efforts to increase access to medicine for societies in need.

America's Greenest Banks

Presented to banks that have implemented distinctive strategies with discernible fiscal and environmental impact, and that present role models and best practice examples for the rest of the industry.

AmeriCares Power of Partnership Award

www.americares.org
    The Power of Partnership award was created by AmeriCares and the Healthcare Distribution Management Association to recognize medical manufacturing and distribution organizations that have shown exceptional dedication to expanding access to medicines and health care for disadvantaged populations in the developing world and in the USA.

Angry Mermaid Award

www.angrymermaid.org
    Established to "recognise the perverse role of corporate lobbyists, and highlight those business groups and companies that have made the greatest effort to sabotage the climate talks, and other climate measures, while promoting, often profitable, false solutions."

Asian Sustainability Rating

www.asiansr.com
    An environmental, social, and governance (ESG) benchmarking tool developed by Responsible Research and CSR Asia. The ASR™ examines the publicly available information of the leading listed companies in ten Asian countries and provides investors, companies, and other stakeholders with a view of their strategic sustainability.

ASSET4 ESG Ratings

www.asset4.com

ASSET4 is a Swiss company that collects ESG data on over 3000 global companies.

B Ratings System (B Corporation)

www.bcorporation.net

The B Impact Rating System is a useful management tool to assess companies' impact on each of their stakeholders and to improve their social and environmental performance using the tools and best practices embedded in the survey.

Best Employers for Workers Over 50

www.aarp.org/work

An AARP rating system [previously American Association of Retired Persons; now a more general 50+ members' organization].

Best German Sustainability Report

www.basf.com

This ranking assesses the transparency, completeness, and thus reliability of the information published by Germany's 150 largest companies on sustainability issues such as environmental protection and employer responsibility.

Best Workplaces for Commuters

www.bestworkplaces.org

Best Workplaces for Commuters is an innovative membership program that provides qualified employers with national recognition and an elite designation for offering outstanding commuter benefits, such as free or low-cost bus passes and vanpool fares, and strong telework programs. Employers that meet the National Standard of Excellence in commuter benefits – a standard created by the National Center for Transit Research and the US Environmental Protection Agency (EPA) – can get on the list of Best Workplaces for Commuters and receive high-level programs and services.

Boston College Center for Corporate Citizenship and Reputation Institute's CSR Index

http://blogs.bcccc.net/2008/10/perception-of-corporate-responsibility-linked-to-reputation

A ranking of the top fifty companies in the USA that the public distinguishes for corporate social responsibility (CSR). Johnson & Johnson, the Walt Disney Company, and Kraft Foods topped the 2010 CSR Index.

Brand Keys Customer Loyalty Engagement Index

www.brandkeys.com/awards

The Brand Keys data paint a detailed picture of the category drivers that engage customers, engender loyalty, and drive real profits. These drivers not only define how the consumer will view the category, compare offerings, and, ultimately, buy, but also identify the expectations the consumer holds for each driver.

Britain's Most Admired Companies

www.bmac.managementtoday.com/BMAC_HowItWorks.htm

Management Today's Britain's Most Admired Companies awards offer a unique insight into the real factors behind corporate reputation. Winners are identified by peer review: Britain's top companies and their bosses are asked to assess their rivals, a revealing exercise that really gets to the heart of what makes businesses succeed.

Building Public Trust Awards

www.bptawards.com

Presented annually by PricewaterhouseCoopers UK, the awards recognize three organizations for their excellence in corporate reporting, one from each of the FTSE 100, FTSE 250, and the public sector (in association with the National Audit Office).

Business in the Community (BITC) CommunityMark

www.bitc.org.uk/community/communitymark/communitymark_companies/index.html

Forty-one companies have achieved the CommunityMark since its launch. All CommunityMark achievers have been recognized for demonstrating excellence in their holistic and strategic approach to community investment.

Business in the Community (BITC) CR Index

www.bitc.org.uk/about_bitc/index.html

A business-led charity with a growing membership of 850 companies, from large multinational household names to small local businesses and public-sector organizations, that advises, supports, and challenges its members to create a sustainable future for people and the planet and to improve business performance.

Carbon Disclosure Project (CDP) Leadership Index

https://www.cdproject.net/en-US/Pages/HomePage.aspx

The only global climate-change reporting system. "Climate change is not a problem that exists within national boundaries. That is why we harmonize climate change data from organizations around the world and develop international carbon reporting standards."

Ceres Water Risk Benchmark

www.ceres.org

Ceres helps members of the Investor Network on Climate Risk leverage their power as shareholders to secure meaningful commitments on sustainability challenges companies are facing. Through shareholder resolutions and face-to-face meetings, investors working with Ceres are asking companies to boost their focus and transparency on sustainability-related risks and set tangible performance goals for mitigating those risks – whether by lowering carbon emissions, reducing water and energy use, or developing more sustainable technologies and products.

Ceres–ACCA Sustainability Reporting Awards

www.ceres.org

Ceres and the Association of Chartered Certified Accountants give awards for the best sustainability reports in North America, presented each year at the Ceres Conference. The purpose of the awards program is to acknowledge and publicize best practice in reporting on sustainability and environmental and social performance by corporations and organizations.

Climate Counts Company Scorecards

www.climatecounts.org

"We score the world's largest companies on their climate impact to spur corporate climate responsibility and conscious consumption. Our goal is to motivate deeper awareness among consumers – that the issue of climate change demands their attention, and that they have the power to support companies that take climate change seriously and avoid those that don't."

CO_2 Benchmark

www.co2benchmark.com

Carbon accounting and metrics are still in the early stages of development. CO_2 Benchmark uses the quartiles approach to set the right level of granularity for comparison. This approach splits the spectrum of performances into four ranges, each capturing 25% of the companies. The first quartile, for example, represents the range of the top 25% performances.

Communitas Awards

www.communitasawards.com

Communitas Awards seeks to honor those special companies, organizations, and individuals that go beyond rhetoric and whose commitment sets them apart from their competition. Communitas winners are dedicated to helping the less fortunate in their communities and are changing the way they do business to benefit their employees, communities, and environment.

Corporate Knights CSR Rankings for Canadian Companies

www.corporateknights.ca

Corporate Knights Inc. is an independent Canadian-based media company that publishes the world's largest circulation magazine with an explicit focus on corporate responsibility. The mission of Corporate Knights Inc. is to humanize the marketplace. Corporate Knights also publishes the annual Best 50 Corporate Citizens in Canada as a *Globe and Mail* insert, and the annual Global 100 Most Sustainable Corporations in the World, announced each year at the World Economic Forum in Davos.

Corporate Responsibility Index (Australia)

www.corporate-responsibility.com.au

The Corporate Responsibility Index provides a comprehensive web-based survey questionnaire and guidance notes that can assist your company to measure, manage, and report on its corporate responsibility performance in a way that is integrated with the business strategy and existing reporting requirements.

Corporate Sustainability Index Benchmark Report
(Technology Business Research)

www.tbri.com

"TBR provides the valuable perspectives you need to make the difficult strategic business decisions that keep you and your company competitive. With TBR as an independent research partner, you'll always have the most up-to-date, clear, unbiased, and focused information available on the companies, markets, and technologies that shape our industry. This includes gaining knowledge about companies' business strategies, their methods of implementation, their strengths and weaknesses, products, positioning, and many other critical business insights."

CorporateRegister.com Reporting Awards

www.corporateregister.com

"CorporateRegister.com is the primary reference point for corporate responsibility (CR) reports and resources worldwide. The majority of our content is available free of charge as a service to the global CR stakeholder community. We also provide specialist tools for advanced users where we identify a need – some of these may be fee based. We strive to maintain the quality of our site and services, in order to advance CR globally."

Covalence EthicalQuote Ranking

www.covalence.ch

The Covalence EthicalQuote Ranking allows corporations to monitor their ethical reputation, to benchmark against peers, and to communicate internally and externally.

CR Magazine 100 Best Corporate Citizens

http://thecro.com

CR: Corporate Responsibility Magazine's 100 Best Corporate Citizens List is known as the world's top corporate responsibility ranking based on publicly available information, and recognized by *PR Week* as one of America's top three most important business rankings.

CRD Analytics: Global Sustainability Index 50

www.crdanalytics.com

"CRD Analytics is the leading provider of independent sustainability investment analytics. Using its proprietary SmartView™ Technology Platform, CRD Analytics empowers its clients with actionable and performance-driven information distilled from large sets of complex data including financial, ESG and patent information. CRD Analytics partners with its clients to construct proprietary index-based products: exchange-traded funds, separately managed accounts, mutual funds, and unit investment trusts."

CSR Survey of Hang Seng Index

www.csr-asia.com/service.php

"CSR Asia can guide you through the entire reporting process and provide you with credible tools for reporting to your key stakeholders. We can also help to design a communication system that is useful in business planning, analysis and management strategy."

CSRHUB Ratings

www.csrhub.com

CSRHUB gives users the information they need to evaluate corporate social values and sustainability. "Once armed with ratings based on their own values, users can take action to both change their behavior and to change the world."

DiversityInc's Top 50 Companies for Diversity

www.diversityinc.com

The DiversityInc Top 50 list is derived exclusively from corporate survey submissions. Companies are evaluated within the context of their own industry, with more than fifteen industries represented.

Diversum Ratings

www.diversum.net

The diversum finance label is open to all organizations eager to promote cultural diversity. It is awarded for a one-year period, subject to the result of checks performed by the diversum association.

Dow Jones Sustainability Indices

www.sustainability-index.com

The Dow Jones Sustainability Indices are the first global indices tracking the financial performance of the leading sustainability-driven companies worldwide. Based on the cooperation of Dow Jones Indices and SAM Sustainability Investing, they provide asset managers with reliable and objective benchmarks to manage sustainability portfolios.

EIRIS Country Sustainability Ratings/Profiles

www.eiris.org

The EIRIS Country Sustainability Rating tool enables investors to create responsible sovereign bond portfolios. Countries can be compared, ranked, and filtered using a wide range of ESG indicators.

Ethibel Sustainable Indices

www.ethibel.org

The Ethibel Sustainability Index provides a comprehensive perspective on the financial performance of the world's leading companies in terms of sustainability for institutional investors, asset managers, banks, and retail investors.

Ethical Corporation Awards

http://events.ethicalcorp.com

Provides business intelligence for sustainability. "We publish the leading global responsible business magazine and independent research, and annually honor leading companies and individuals for their contributions to sustainability."

Ethisphere World's Most Ethical Companies

http://ethisphere.com

"Ethisphere is an influential beacon for both good companies, aspirants and laggards alike. Ethisphere uniquely recognizes that even companies with less than perfect historical corporate citizenship operating records need to have their successes celebrated along the way if they decide to conscientiously improve their ethics, compliance and community practices. Anything else would simply build those companies' resistance to productive engagement with third parties. It's this sort of pragmatic approach that actually delivers Ethisphere's power to influence, and help make and retain these companies as productive members of the corporate citizenship community."

Forbes' 100 Most Trustworthy Companies

www.forbes.com/2010/04/05/most-trustworthy-companies-leadership-governance-100.html

The American businesses that have the most transparent and conservative accounting practices and most prudent management.

Forest Footprint Disclosure

www.forestdisclosure.com

Forest Footprint Disclosure (FFD) is a special project of the Global Canopy Foundation, initiated in 2008. FFD engages with private-sector companies to ask them to disclose their current understanding of their "forest footprint" based on exposure to five key commodities – soy, palm oil, timber, cattle products, and biofuels – in their operations and/or their supply chains. All these commodities have the potential to be sourced from recently deforested land. FFD goes beyond simple disclosure and acts as a catalyst for investors to really engage with the companies in their portfolios on this issue.

FT Sustainable Banking Awards

www.ftconferences.com

"The FT/IFC Sustainable Finance Awards highlight how financial institutions create shared value that benefits both shareholders and society [...] This year we are also recognising how sustainable finance, notably pension funds and private equity, is supporting small and medium-sized enterprises, a key source of job creation and economic growth in most countries."

FTSE CDP Carbon Strategy Index Series

www.ftse.com

The FTSE CDP Carbon Strategy Index Series are carbon-risk-tilted versions of FTSE's established benchmark indices, and have been developed in partnership with the world's foremost experts in carbon data collection, management, and analysis.

Global 100 Most Sustainable Corporations in the World

www.global100.org

"The Global 100 Most Sustainable Corporations in the World is an annual project initiated by *Corporate Knights*, the magazine for clean capitalism. In 2010, *Corporate Knights* collaborated with three strategic partners to identify the *Corporate Knights* Global 100 Most Sustainable Corporations in the World. The *Corporate Knights* Global 100 team included Inflection Point Capital Management […] Legg Mason's Global Currents Investment Management, and Phoenix Global Advisors LLC (a consulting and technology platform focused on sustainability). As with past editions of the Global 100, the aim was to highlight the global corporations which have been most proactive in managing ESG issues."

Globe Award for Sustainability Reporting

www.globeaward.org

Globe Award is a world-covering sustainability award which is given out in four categories to recognize and encourage societies, the corporate sector, individuals, and academia that have excelled in the area of sustainability. Globe Award is a not-for-profit organization with the aim of fostering sustainable development in society. "Our role is to inspire actors to do more and to take responsibility through awarding great cases within research, innovations, large corporations and cities.

GMI Company Ratings (GovernanceMetrics International)

www.gmiratings.com

GMI uses a proprietary scoring algorithm to produce its ratings. The GMI research template is divided into six broad categories of analysis. These categories are further divided into sub-sections. Each individual metric has a numerical value and each sub-section and research category is weighted according to investor interest.

Golden Peacock Awards

www.goldenpeacockawards.com

"The Golden Peacock Awards are now recognized worldwide as the hallmark of corporate excellence because of their independence, integrity, transparency and through evaluation of application provides opportunity for self assessment and helps companies accelerate the performance and competition. All institutions whether public, private, non-profit, government, business, manufacturing and service sector are eligible to apply. Leadership Awards are determined through nomination. They provide not only worldwide recognition and prestige but also a competitive advantage in driving business in this tumultuous world under Golden Peacock Award models. The Awards are bestowed annually and are designed to encourage total improvement in each sector of our business."

Goodness 500

http://goodness500.org

Goodness 500 is a social enterprise that helps people assess the social responsibility of the companies with the most power to change the world.

Green Awards for Creativity in Sustainability

www.greenawards.com

By showcasing "best in class" examples of effective, innovative, and creative approaches to sustainability, the International Green Awards™ aims to be an agent of change. "To this end, the Awards team searches the globe every year to find true influencers, leaders, entrepreneurs, and innovators, identifying worldwide sustainability success stories that inspire and motivate others. It is no exercise in fluffy back-patting."

Green Effie Awards

www.effie.org

The Effie network has teamed up with top research and media organizations worldwide to bring its audience the most relevant and first-class insights into effective marketing strategy. This has taken shape in an annual webinar series featuring in-depth, integrated case studies co-hosted in partnership with Google, research from a growing partnership with New York University, student competitions allowing Effie to educate and engage with future industry leaders, global conferences to share the insights and best practices of its 40+ programs worldwide, winners' showcases, and more.

Greenopia Brand and Product Ratings

www.greenopia.com

"Greenopia is your local guide to green living. Our mission is a basic and big one: We set out to create a directory of eco-friendly retailers, services, and organizations and conducted extensive research on those we listed in the guide. This guide is not a paid directory; companies cannot pay to be included and all listees are included because they met our strict standards of eco-friendliness. They have already been screened for their sustainability in the product or service arena and are now being compared with 'the best of the best.'"

Greenpeace Cool IT Challenge Leaderboard

www.greenpeace.org/international/Global/international/publications/
climate/2010/Cool%20IT%20V3%20full%20report%282%29.pdf

"Greenpeace's Cool IT Challenge calls on leading information technology (IT) companies to be champions in the fight to stop climate change. The IT sector possesses the innovative spirit, technological know-how, and political influence to bring about a rapid clean energy revolution. The IT industry must boldly step out in front of older, entrenched energy companies to develop a robust business model that helps the world achieve critical emissions reductions – a win–win for the IT industry as it pioneers a global shift to a clean energy economy."

GRI Readers' Choice Awards

www.globalreporting.org

"The Global Reporting Initiative (GRI) is a network-based organization that pioneered the world's most widely used sustainability reporting framework. GRI is committed to the Framework's continuous improvement and application worldwide. GRI's core goals include the mainstreaming of disclosure on ESG performance. GRI's Reporting Framework is developed through a consensus-seeking, multi-stakeholder process. Participants are drawn from global business, civil society, labor, academic and professional institutions."

GS SUSTAIN Focus List

www2.goldmansachs.com

GS SUSTAIN is a unique global equity strategy that brings together ESG criteria, broad industry analysis and return on capital to identify long-term investment opportunities.

Guide to Greener Electronics

www.greenpeace.org/international/en/campaigns/toxics/electronics/Guide-to-Greener-Electronics

Greenpeace guide that ranks the eighteen top manufacturers of personal computers, mobile phones, TVs, and games consoles according to their policies on toxic chemicals, recycling, and climate change.

HIP 100 Index

www.hipinvestor.com

The HIP Scorecard ratings and rankings – which focus on quantifiable metrics of sustainability and how they drive financial performance – have been featured in *Fast Company* magazine and online in 2007, 2008, and 2010. HIP has also advised corporate clients, such as Walmart and Nike, on how to be more HIP ("HIP = human impact + profit = the new fundamentals of investing").

InfoWorld Green 15 Awards

www.infoworld.com

"InfoWorld provides in-depth technical analysis on key products, solutions, and technologies for sound buying decisions and business gain. InfoWorld is the place to turn for in-depth analysis of the issues, trends, and products that run your enterprise. InfoWorld features trusted industry columnists, a sharp focus on IT issues, and product test results and reviews backed by the renowned InfoWorld Test Center."

Inrate Sustainability Assessments

www.inrate.com

Inrate is an independent sustainability rating agency based in Switzerland. "Since 1990, we have been linking our sound understanding of sustainability with innovative research solutions for the financial markets – the main reason why we belong to the largest and most respected agencies in Europe."

Jantzi Social Index

www.sustainalytics.com/jantzi-social-index-may-2011-returns

Jantzi-Sustainalytics reported that the Jantzi Social Index® (JSI) decreased in value by 0.71% during May 2011. During the same period, the S&P/TSX Composite Index decreased by 0.87% and the S&P/TSX 60 Index decreased by 0.97%. Since inception on January 1, 2000 through May 31, 2011, the JSI has achieved an annualized return of 6.47%, while the S&P/TSX Composite and the S&P/TSX 60 had annualized returns of 6.73 and 6.37%, respectively, over the same period.

Management & Excellence Rankings

www.management-rating.com

Management & Excellence (M&E) is a research and rating company in the areas of ethics, sustainability, corporate governance, transparency, and CSR specializing in Latin America, Spain, and the oil industry worldwide.

Maplecroft Climate Innovation Indices

http://maplecroft.com/cii

"Our award-winning services empower responsible organizations to identify, manage and mitigate global risks in their operations, supply chains and distribution networks. Maplecroft is a leader in human rights due diligence, impact assessment and monitoring. We have longstanding experience in ethical supply chain management with the technological capability to custom build monitoring and reporting systems that can merge our risk intelligence with company data feeds."

MSCI ESG Indices

www.msci.com

"MSCI is a leading provider of investment decision support tools to over 5,000 clients worldwide, ranging from large pension plans to boutique hedge funds. We offer a range of products and services – including indices, portfolio risk and performance analytics, and governance tools – from a number of internationally recognized brands such as Barra, RiskMetrics and ISS."

NASDAQ OMX CRD Global Sustainability 50

https://indexes.nasdaqomx.com

The NASDAQ OMX CRD Global Sustainability Index is an equally weighted equity index that serves as a benchmark for stocks of companies that are taking a leadership role in sustainability performance reporting and are traded on a major US stock exchange. The Index is made up of companies that have taken a leadership role in disclosing their carbon footprint, energy usage, water consumption, hazardous and non-hazardous waste, employee safety, workforce diversity, management composition, and community investing.

Newsweek *Green Rankings*

www.newsweek.com

Newsweek's 2010 Green Rankings is a data-driven assessment of the largest companies in the USA and in the world. "Our goal was to cut through the green chatter and quantify the actual environmental footprints, policies, and reputations of these big businesses. To do this, we teamed up with three leading environmental research organizations to create the most comprehensive rankings available."

Oceana's Grocery Store Guide

www.oceana.org

Oceana is the largest international organization working solely to protect the world's oceans. "Oceana wins policy victories for the oceans using science-based campaigns. Since 2001, we have protected over 1.2 million square miles of ocean and innumerable sea turtles, sharks, dolphins and other sea creatures. More than 500,000 supporters have already joined Oceana. Global in scope, Oceana has offices in North, South and Central America and Europe."

Oekom Corporate Ratings

www.oekom-research.com/index_en.php

"Oekom research is one of the leading rating agencies worldwide in the field of sustainable investment. We analyse companies and countries with regard to their environmental and social performance."

OMX GES Ethical Index Series

www.ges-invest.com

The OMX GES Sustainability Finland Index is a benchmark index, which comprises the forty leading Finnish listed companies in terms of sustainability. The index criteria are based on international guidelines for ESG issues and support investor considerations to the UN Principles for Responsible Investment. The assessment is conducted by GES Investment Services, which rates the most traded companies on the NASDAQ OMX Helsinki exchange on their preparedness on ESG issues.

P&G Supplier Environmental Sustainability Scorecard

www.pg.com

"Our market capitalization is greater than the GDP of many countries, and we serve consumers in more than 180 countries. With this stature comes both responsibility and opportunity. Our responsibility is to be an ethical corporate citizen – but our opportunity is something far greater, and is embodied in our Purpose."

PR News *CSR Awards*

www.prnewsonline.com

"The *PR News* awards programs provide you the opportunity to showcase your best communications initiatives of the year and salute outstanding performance by individuals. We have nine elite award programs that bring to light top talent and creativity across the spectrum of the public relations and communications arena."

RepRisk Index

www.reprisk.com

RepRisk monitors companies and projects ESG risk exposure. In particular, RepRisk helps with the identification of controversial companies and projects in order to avoid financial, compliance, and reputational risks.

RepuTex Sustainability / ESG Ratings

www.reputex.com

"RepuTex is a leading research and advisory firm, specializing in carbon risk analytics for global companies and investment professionals – it is the foremost knowledge source for market participants seeking to measure, benchmark and price carbon risk and exposure."

S&P ESG India

www.standardandpoors.com

The S&P ESG India index provides investors with exposure to fifty of the best performing stocks in the Indian market as measured by ESG parameters. "The index represents the first of its kind to measure environmental, social, and corporate governing practices based on quantitative rather than subjective factors with the implementation of a unique and innovative methodology."

S&P Shariah Indices

www2.standardandpoors.com/spf/pdf/index/Shariah_Methodology.pdf

"In 2006, Standard & Poor's introduced the S&P Shariah indices. Shariah is Islamic canonical law, which observant Muslims adhere to in their daily lives. Shariah has certain strictures regarding finance and commercial activities permitted for Muslims. Over the last few years, the demand for Shariah compliant financial products has increased. Standard & Poor's has contracted with Ratings Intelligence Partners (RI) to provide the Shariah screens and filter the stocks based on these screens. RI is a London/Kuwait-based consulting company specializing in solutions for the global Islamic investment market. Its team consists of qualified Islamic researchers who work directly with a Shariah Supervisory Board. It is continually working with regional banks to create Shariah-compliant equity products and expand investment offerings."

Scrip Awards

www.scripintelligence.com/awards

"The Scrip Awards provides the industry an opportunity to acknowledge and applaud its highest achievers across all parts of the value chain, and recognize both corporate and individual achievement. Entries to the awards are nominated by you, the biopharmaceutical industry. A panel of highly respected and independent luminaries from across the sector then produces a shortlist of nominees and chooses the winners for each category. Looking at the reaction to previous winners we know how highly regarded these awards are in the industry. It is truly a great honour to be on the shortlist of nominees. But remember you have to be in it to win it. Please feel free to nominate candidates for as many of the categories as you wish. Self-nomination is not only permitted it is actively encouraged."

Tomorrow's Value Rating

www.tomorrowsvaluerating.com

"The Tomorrow's Value Rating (TVR) is an annual assessment of corporate sustainability practices among leading companies worldwide. Run by international corporate sustainability agency Two Tomorrows, the TVR aims to further the debate on sustainable business, identify and reward best practices, and spur healthy competition among companies, motivating them to create tomorrow's value."

Toxic 100 Air Polluters

www.peri.umass.edu/toxic100

The Toxic 100 Air Polluters index identifies the top US air polluters among the world's largest corporations. The index relies on the US Environmental Protection Agency's Risk Screening Environmental Indicators (RSEI), which assess the chronic human health risk from industrial toxic releases. The underlying data for RSEI is the EPA's Toxics Release Inventory, in which facilities across the USA report their releases of toxic chemicals. In addition to the amount of toxic chemicals released, RSEI also includes the degree of toxicity and population exposure. The Toxic 100 Air Polluters ranks corporations based on the chronic human health risk from all of their US polluting facilities.

Trucost Corporate Environmental Data and Profiles

www.trucost.com

"Trucost data enables organizations to identify, measure and manage the environmental risk associated with their operations, supply chains and investment portfolios. Key to our approach is that we not only quantify environmental risks, but we also put a price on them, helping organizations understand environmental risk in business terms."

True Sustainability Index

www.sustainableorganizations.org/true-sustainability-index.html

"The True Sustainability Index™ is the world's first context-based triple bottom line model for measuring, rating, ranking and reporting the true sustainability performance of organizations. It is context-based in the sense that it expresses organizational performance in terms of impacts relative to actual social and environmental conditions in the world, and not just in terms of top line trends, efficiency or incremental effects."

US Chamber of Commerce Business Civic Leadership Center Corporate Citizenship Awards

http://bclc.uschamber.com

The Business Civic Leadership Center addresses social issues that affect business, including CSR, philanthropy, non-profit and social service effectiveness, globalization, community investment, and disaster assistance.

Vaccine Industry Excellence Awards

www.terrapinn.com/2011/vaccine-industry-excellence-awards

The Vaccine Industry Excellence Asia Awards 2011 were created to recognize and celebrate the leaders, innovators, and pioneers in the Asian vaccines industry.

Vigeo Ratings

www.vigeo.com

"Vigeo was founded in 2002 [...] and has established itself as the leading European expert in the assessment of companies and organizations with regard to their practices and performance on ESG issues."

Walmart Sustainability Index

http://walmartstores.com/Sustainability/8844.aspx

"A high-definition journey through four continents, exploring personal stories of breakthrough change in every corner of the Walmart world. A kaleidoscope of top sustainability experts, CEOs, suppliers and entrepreneurs show how companies are doing well by doing good – tackling sustainability head on to out-compete peers and lead true change in the process."

World Environment Center Gold Medal for International Corporate Achievement in Sustainable Development

www.wec.org

"World Environment Center (WEC) is a global non-profit, non-advocacy organization that advances sustainable development through the business practices of member companies and in partnership with governments, multi-lateral organizations, non-governmental organizations, and other stakeholders. The WEC Gold Medal Award is presented annually to a global company that has demonstrated a unique example of sustainability in business practice and is one of the most prestigious forms of recognition of a global company's ongoing commitment to the practice of sustainable development."

World's Top Sustainable Stocks (SB20)

www.sustainablebusiness.com

"Every person and organization must decide the path they will take to bring this vision to reality. At SustainableBusiness.com, we serve the businesses that are ready to make this vision a reality now. We provide global news and networking services to help green business grow. Rather than covering a slice of the industry, we offer visitors a unique lens on the field as a whole, covering all sectors that impact sustainability: renewable energy/efficiency, green building, green investing, and organics."

Index